mechanical technology in ENGLISH

알기 쉬운 기계기술영어

정태환 저

普文堂

머리말

오늘날 무수히 많은 직업 중 어느 분야에 종사하든 영어를 모르고서는 한 발자국도 전진할 수 없는 시대가 되었음은 누구나 인식하고 있는 바이다. 특히 하루가 다르게 변하고 있는 첨단 과학기술을 습득해야 하는 공학도 들에게는 영어로 된 전공서적의 독해능력이 성공여부와 직결됨은 너무나 자명한 사실이라고 하겠다.

그러나 대학생들이 전공영어를 공부하려는 의욕은 있으나 마땅한 책이 없었다. 과거에 출간된 몇 종의 전공영어 교재들은 내용이나 교재면에서 학습에 많은 어려움을 안고 있다.

이러한 점을 감안하여 이 공업영어 책을 출간함에 있어서 기존 서적들의 틀을 벗어나 비교적 적은 분량으로 최단 시간에 독해능력을 해결할 수 있음은 물론 기계 각 분야의 핵심지식을 습득할 수 있도록 즉, 1석 2조의 효과를 얻는데 최선을 다해 정성을 들였다.

아무쪼록 이 책이 기계분야를 공부하는 모든 대학인들에게 좋은 반려가 되어 학교성적의 향상은 물론 전공분야에 대한 실력이 길러져서 졸업 후 향방에 알찬 결실이 이루어지기를 기대하며 또 그 기대에 부응할 수 있으리라고 확신한다.

끝으로 이 책이 출간될 수 있도록 물심양면으로 후원하여 주신 보문당 출판사 여러분들께 깊은 감사드린다.

2000년 2월

저자 씀

CONTENTS

Chapter 1 HAND TOOLS / 11

1.1 Screwdrivers / *11*
1.2 Wrenches / *12*
1.3 Pliers / *17*
1.4 Hammers / *18*
1.5 Chisels / *19*
1.6 Files / *20*
1.7 Punches / *21*
1.8 Hacksaw / *22*

Chapter 2 MEASUREMENTS AND MEASURING TOOLS / 25

2.1 Ruler / *25*
2.2 Feeler Gages / *26*
2.3 Wire Feeler Gages / *26*
2.4 Calipers / *27*
2.5 Micrometer / *28*
2.6 Vernier Caliper / *29*

CONTENTS

Chapter 3 TECHNICAL DRAWING / 33

3.1 Types Of Lines / 33
3.2 Application Of Types Of Lines / 34
3.3 First-Angle And Third-Angle / 35
3.4 First-Angle Projection / 36
3.5 Third-Angle Projection / 37
3.6 Sectioning / 37
3.7 Symbols For Section Lining / 48
3.8 Tolerance / 48
3.9 Tolerance Display System / 49
3.10 Classes Of Fits / 50

Chapter 4 CHOOSING A MATERIAL / 53

4.1 Introduction / 53
4.2 Mechanical Properties / 54
4.3 Engineering Materials / 59
4.4 Ferrous Metals / 59
4.5 Non-Ferrous Metals / 61
4.6 Plastics Materials / 65
4.7 Selecting The Materials / 67

CONTENTS

Chapter 5 MANUFACTURING TECHNOLOGY / 71

5.1 Introduction / 71
5.2 Casting / 71
5.3 Rolling / 75
5.4 Pressing (Sheet-Metal Forming) / 78
5.5 Forging / 83
5.6 Plastics Moulding / 84
5.7 Machining / 86

Chapter 6 WELDING / 93

6.1 Introduction / 93
6.2 Oxy-Acetylene Gas Welding / 94
6.3 Arc Welding / 97
6.4 Resistance Welding / 100

Chapter 7 SAFETY IN THE SHOP / 103

7.1 Accidents / 103
7.2 Protective Clothing / 104
7.3 Head Protection / 107
7.4 Eye Protection / 108
7.5 Hand Protection / 109
7.6 Foot Protection / 109
7.7 Lifting And Carrying / 110

CONTENTS

Chapter 8 MECHANICAL FASTENERS / 113

8.1 Introduction / 113
8.2 Screw Thread Terminology / 113
8.3 Single And Multiple Thread / 115
8.4 Screw-Thread Forms / 116
8.5 External Thread Specifications / 118
8.6 Internal Thread Specifications / 118
8.7 Common Screw Fasteners / 119
8.8 Lock Washers / 123
8.9 Snap Rings / 124
8.10 Keys / 125
8.11 Pins / 126

Chapter 9 FUNDAMENTALS OF GEAR / 129

9.1 Introduction / 129
9.2 Spur Gear Terminology / 130
9.3 Rack And Pinion / 132
9.4 Internal Gears / 133
9.5 Bevel Gears / 134
9.6 Worm And Wheel / 136
9.7 Gear Trains / 137
9.8 Simple Gear Trains / 137
9.9 Compound Gear Trains / 139

CONTENTS

Chapter 10 STRENGTH OF MATERIALS / 143

10.1 Stress / *143*

10.2 Strain / *143*

10.3 Hook' s Law / *144*

10.4 Modulus Of Elasticity Or Young' s Modulus / *144*

10.5 Stress-Strain Graph For Mild Steel / *145*

10.6 Tensile Strength And Factor Of Safety / *147*

10.7 Shearing / *148*

Chapter 11 FLUID MECHANICS / 151

11.1 Introduction / *151*

11.2 Density And Specific Gravity / *152*

11.3 Buoyant Force / *154*

11.4 Pressure / *155*

11.5 Pressure Increase With Depth / *155*

11.6 Pascal's Principle / *157*

11.7 Hydraulic Press / *157*

11.8 Hydraulic Lift / *158*

Chapter 12 INTERNAL COMBUSTION ENGINE / 163

12.1 Introduction / *163*

12.2 Four-Stroke Cycle Engine / *164*

12.3 Diesel Engines / *165*

12.4 Engine Construction / *165*

CONTENTS

Chapter 13 COMPUTERS / 173

13.1 Introduction / *173*
13.2 Personal Computer Systems / *174*
13.3 Keyboards / *175*
13.4 Monitors / *176*
13.5 Printers / *176*
13.6 Floppy Disks And Disk Drivers / *177*
13.7 Hard Disk Drives / *178*
13.8 CD-ROM Disks / *180*
13.9 Modems / *180*

Chapter 14 TECHNOLOGY OF THE CARS / 183

14.1 Introduction / *183*
14.2 Safety / *184*
14.3 Flashback / *185*
14.4 How A Car Works / *185*
14.5 Construction Of Cars / *186*
14.6 Instruments For The Driver / *189*

Chapter 15 AIRCRAFT / 193

15.1 Introduction / *193*
15.2 Shapes / *194*
15.3 Power / *196*
15.4 The Changing Face Of Design / *196*
15.5 Glossary For Aircraft / *197*

1

Hand Tools Chapter

(손 공 구)

1.1 SCREWDRIVERS

The typical **screwdriver** (Fig. 1.1) has a handle and a shank, the end of which has been flattened to form a blade. The blade tip should be flat on the end so that it fits squarely into the screw slot (Fig. 1.2). If it is tapered, it will tend to rise out of the screw slot when it is turned. This makes the removal or installation of a screw more difficult.

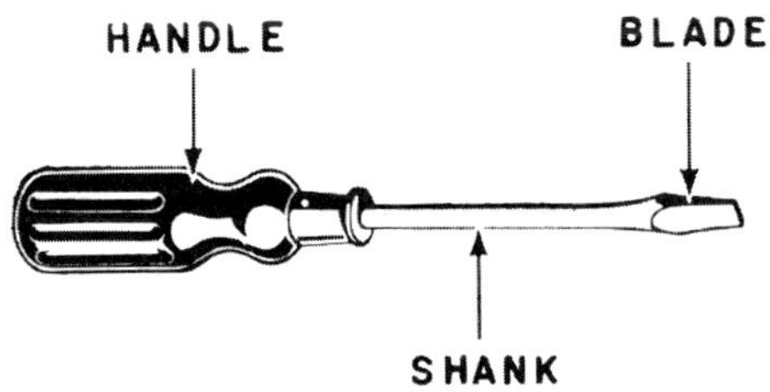

Fig. 1.1 A screwdriver

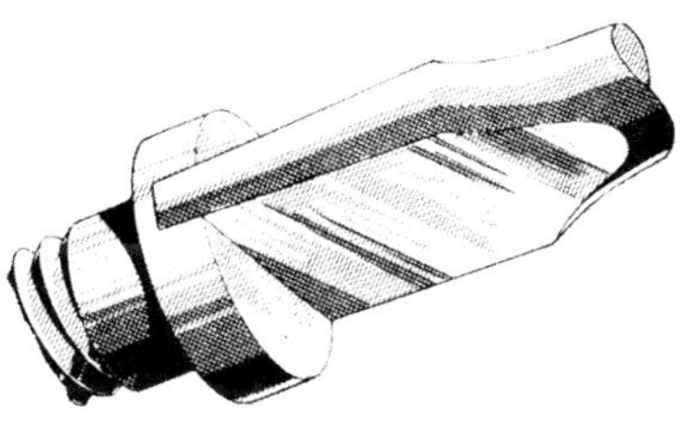

Fig. 1.2 The screwdriver tip should fit the screw slot precisely

Chapter 1

Some screws have what resembles two screw slots at right angles. The **Phillips screw** is one of these. It requires a special screwdriver with a tip to fit the crossed slots (Fig. 1.3).

Fig. **1.3** Phillips screwdriver and the slotted head of a phillips screw

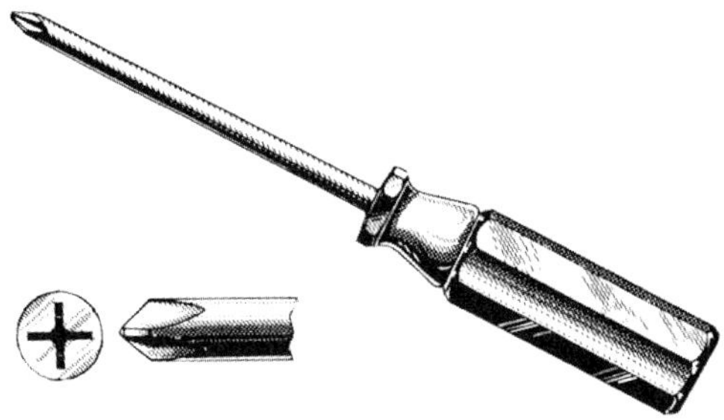

1.2 WRENCHES

Bolt heads and nuts commonly used in machines and engines are of the **hexagonal type.** That is, they have six flats around the outer surface (Fig. 1.4) These flats permit the use of a wrench to turn the nut or bolt.

Fig. **1.4** A hex-head bolt and nut.

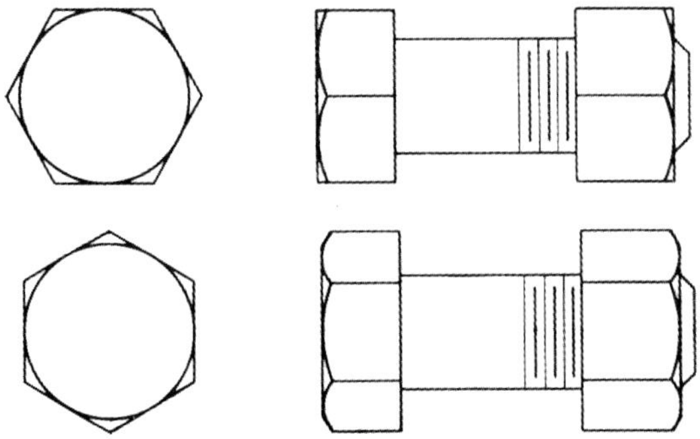

1.2.1 OPEN-END WRENCHES

A typical wrench of the **open-end type** is shown in Fig. 1.5. Most have the open end at an angle of 15° with the handle, as shown, although a few have a 90° angle for special purposes. The open-end wrench is used by placing the open end over the flats of the bolt head or nut. The wrench handle provides considerable leverage so that the bolt or nut can be tightened or loosened easily.

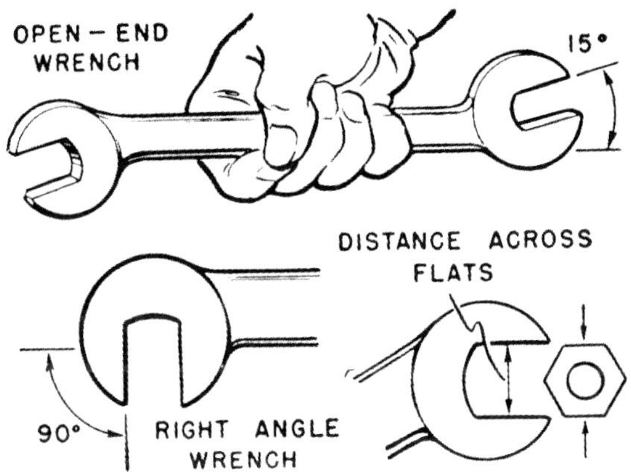

Fig. 1.5 Open-end wrenches

1.2.2 BOX WRENCHES

Box wenches serve the same purpose as open-end wrenches. However, they are more useful in many cases, because the box fits down over the nut or bolt head and will not slip off. The most widely used box wrench is the 12 point, which means the box has 12 notches in it, spaced 30° apart. One disadvantage of the box wrench is that it must be lifted clear off the nut and put on in a different position with each swing of the wrench when working in a restricted space. Here is where the open-end wrench is more convenient to use; it will slip onto the nut more easily. Thus, many wrenches are made as combination open-end and box (Fig. 1.6)

Fig. 1.6 Box wrench and combination wrench

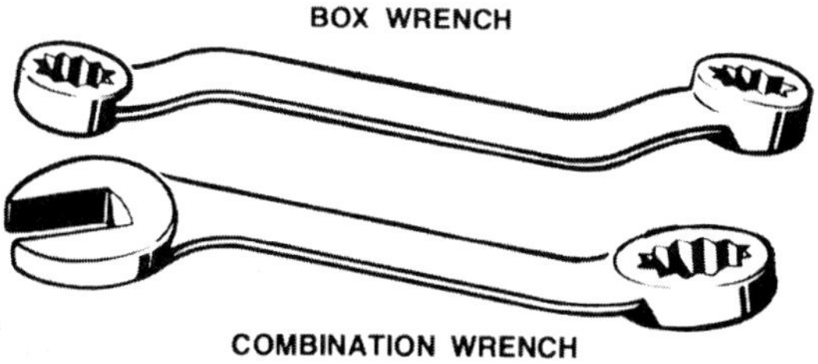

1.2.3 ADJUSTABLE WRENCH

The **adjustable wrench** (Fig. 1.7) has an adjustable jaw that can be moved back and forth to narrow or widen the distance between the jaws. Thus, the adjustable wrench can be made to fit many sizes of nut or bolt heads. This tool is not intended to take the place of the regular wrench but is handy to have for special jobs or odd sizes of nuts or bolts.

Fig. 1.7 Adjustable open-end wrench

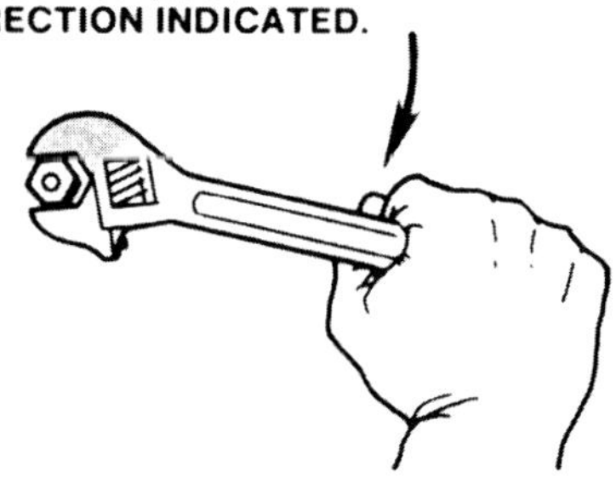

1.2.4 SOCKET WRENCHES

The **socket wrench** is similar to the box wrench, except that the sockets are detachable from the handle so that a single handle can be used for many sizes of sockets. Fig. 1.8 shows two types of sockets. The drive end of a socket fits into the handle, and the socket fits down over the nut or bolt head.

A variety of handles are used with sockets. The ratchet handle permits the handle to be swung back and forth without lifting the socket from the nut or bolt head with each swing (Fig. 1.9).

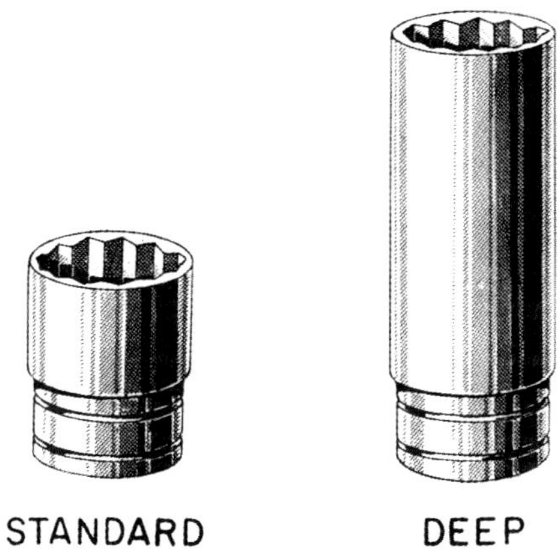

Fig. **1.8** Typical sockets

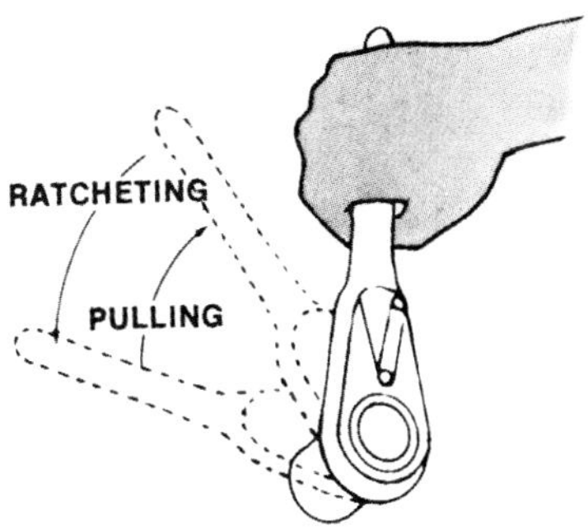

Fig. **1.9** The rachet handle

1.2.5 TORQUE WRENCH

The **torque wrench** (Fig. 1.10) is a special form of wrench that has a measuring device to indicate the amount of torque (twist) being applied to a bolt or nut. In much of today's machinery, engineers have carefully worked out the amount that bolts or nuts should be tightened.

If they are not tightened enough, they may loosen in service. If they are tightened too much, then too much strain may be put on the bolts or machine parts and they might break. In either case, the result could be severe damage to the machine. To prevent this, nuts and bolts must be tightened just the right amount. The torque wrench does this job. As the handle is pulled, the amount of torque is registered on the dial.

Torque wrenches Fig. 1.10

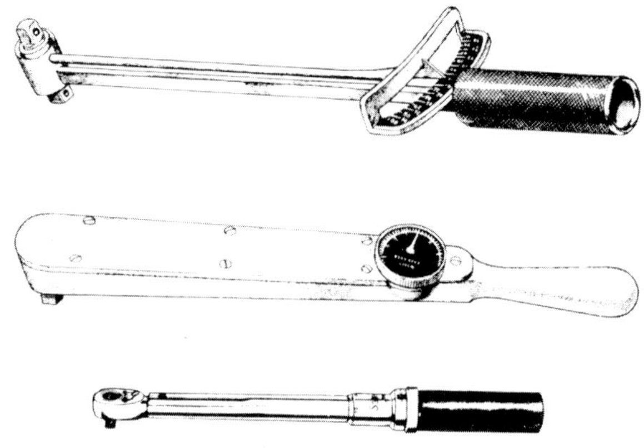

1.3 PLIERS

Pliers (Fig. 1.11) are a special form of adjustable wrench. The jaws are adjustable as the two legs of the handle are moved on the pivot. Thus, the jaws can be widened to grip an object and to hold or turn it. Pliers should not be used on nuts or bolt heads, because they will round off the edges of the hex and roughen the flats so that a wrench will no longer fit properly.

Some pliers have a side cutter that can be used to cut wire. There are also regular **nippers** that have cutting edges instead of jaws as in the pliers. These can be used to cut wire, thin sheet metal, small bolts, and so on.

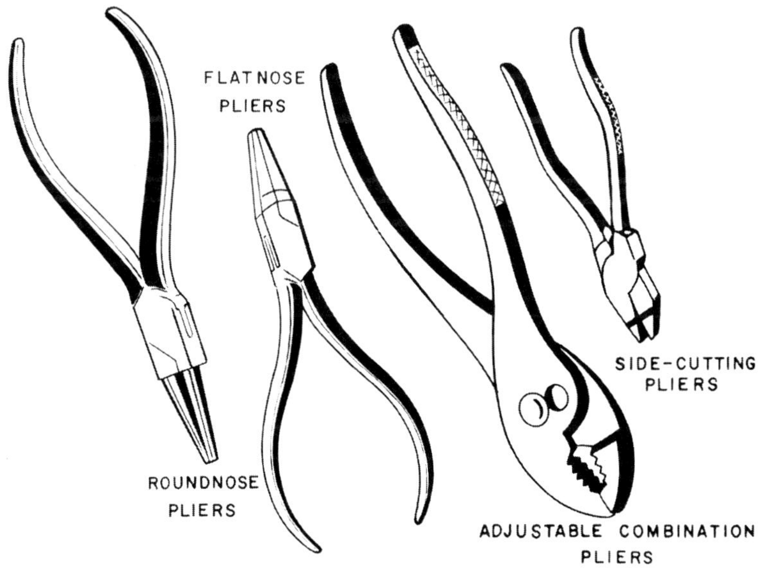

Fig. **1.11** Various types of pliers

1.4 HAMMERS

A variety of **hammers** are used in the shop. You are probably familiar with the claw hammer, which has a claw opposite the driving head for pulling nails. This is a carpenter' s tool and is not of much use around the machine shop. Instead, the **ball peen hammer** is used (Fig. 1.12). When using this hammer, grip the handle firmly near its end and swing it so that the face strikes the object squarely (Fig. 1.13).

Ball peen hammer *Fig.* ***1.12***

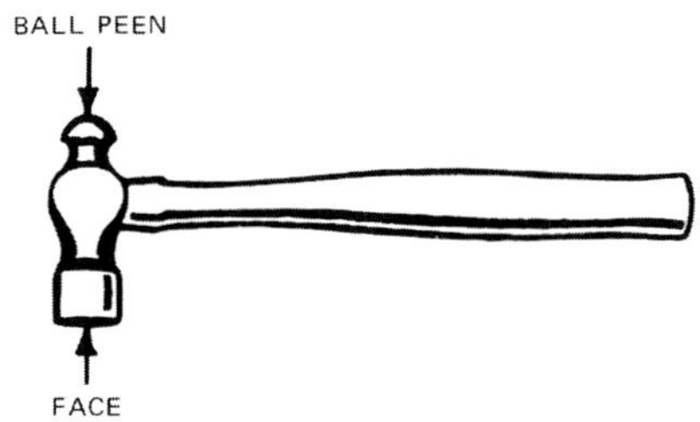

Correct striking action *Fig.* ***1.13***

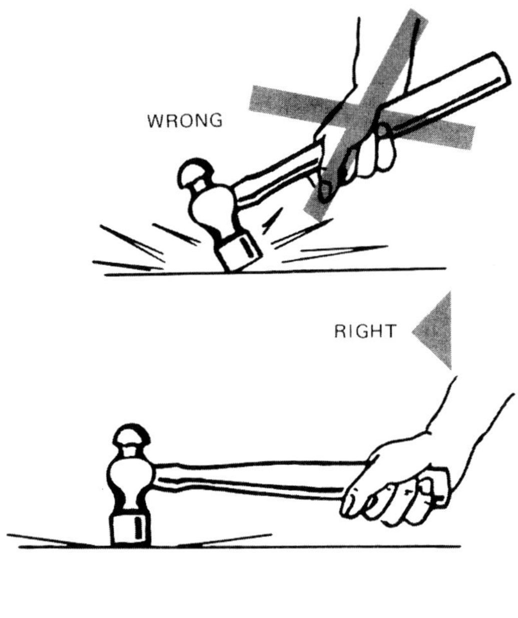

1.5 CHISELS

The **chisel** has a single cutting edge and is driven with a hammer to cut metal. Several different shapes of chisels are shown in Fig.1.14. Each shape has its special purpose, but the chisel most commonly used is the plain flat chisel.

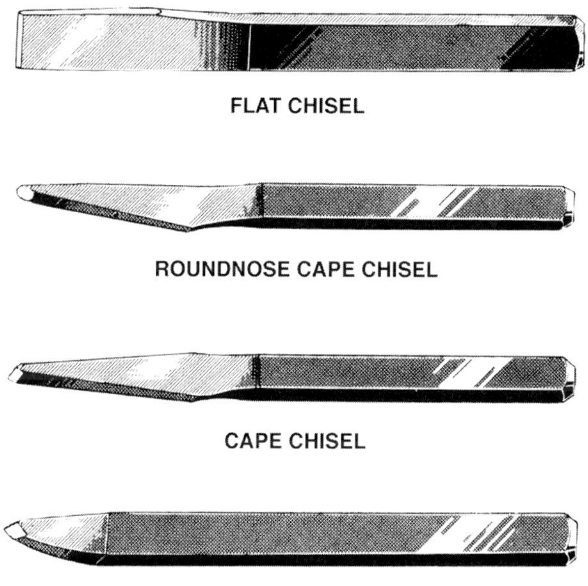

Fig. **1.14** Various types of chisels

Grinding the mushroom from chisel head *Fig.* ***1.15***

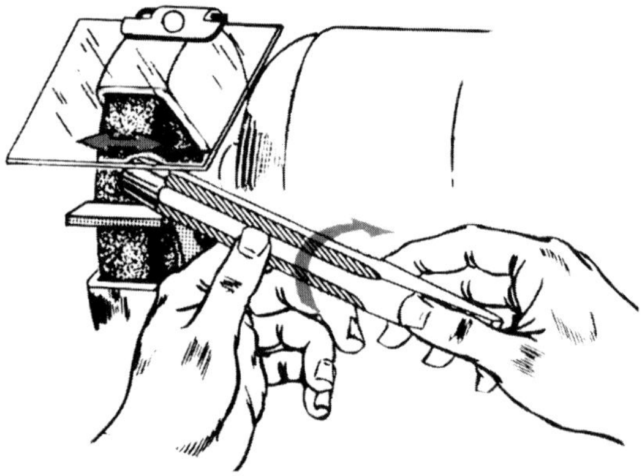

Goggles should be worn when chipping with a chisel, since metal chips may be propelled from the workpiece or chisel and fly into the eyes. A chisel that has mushroomed on the end because of repeated hammer blows should be dressed on a grinding wheel (Fig. 1.15) so that the turned-over metal is removed. Otherwise, this metal could break off and fly away as the hammer strikes it. Not only is it hard to be accurate with a chisel in this condition, but the flying chips might cut your hand or hurt someone else.

1.6 FILES

The **file** is a little like a series of tiny chisels each with a sharp cutting edge. Files have many sizes, shapes, and uses. A typical file with the parts named is shown in Fig. 1.16.

Files are classified according to shape. Files may be **flat, triangular, half-round,** or **round,** and may or may not have taper (Fig. 1.17).

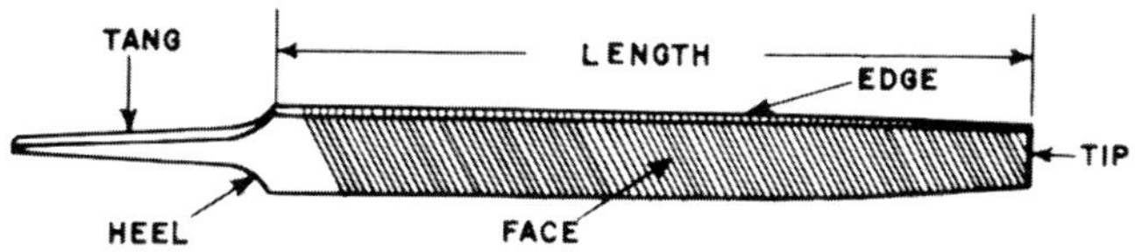

Fig. 1.16 A typical file with parts indicated

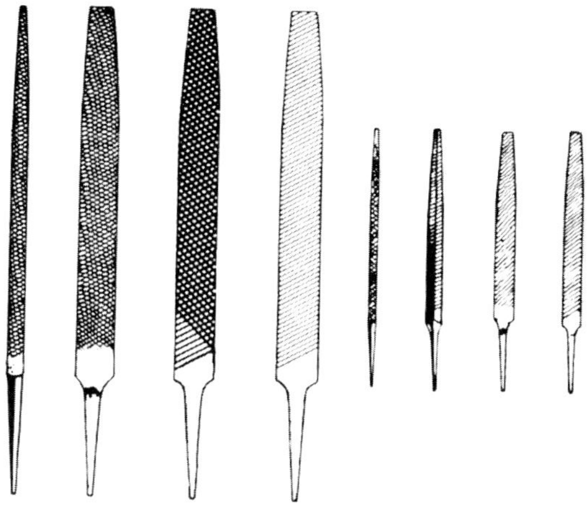

Fig. 1.17 Various types of files

CAUTION

Never attempt to use a file as a pry bar and never hammer on it. The file is brittle and could shatter in a dangerous manner.

1.7 PUNCHES

Punches (Fig. 1.18) are used to knock out rivets or pins and are also used to align parts for assembly and to mark the locations of holes to be drilled. They come in several shapes, each having its own special use.

The **center punch** is used for marking hole locations for drilling.

The **pin punch**, which has a long thin shank, can then be used to drive the pin out.

The **starting punch** is used with a hammer to loosen a pin or shaft.

Various types of punches *Fig.* ***1.18***

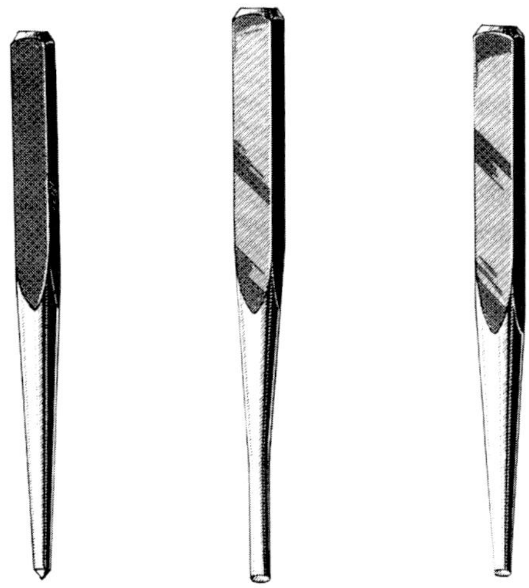

1.8 HACKSAW

The **hacksaw** is the saw commonly used for cutting metals. The blades are replaceable and are held, for sawing, in a metal adjustable frame (Fig. 1.19).

Blades are made with from 14 to 32 teeth per inch. Using a blade with the wrong number of teeth for the job will make the job more difficult and could also cause the blade to break.

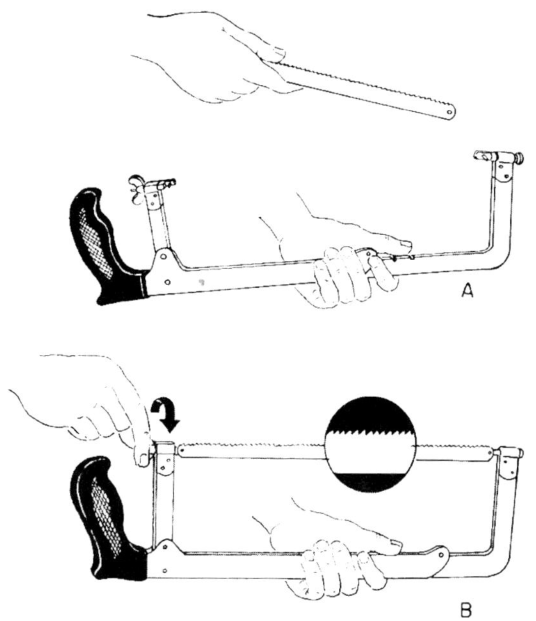

1.19 (A) Installing the blade on a hacksaw.
(B) Tightening the blade. Note that teeth point away from the handle

Chapter 1

VOCABULARY

shank ; (연장의)자루
flatten ; 평평하게 하다. 평평해지다.
blade ; 칼날
taper ; 끝이 점점 가늘어지다.
removal ; 이동, 이전, 제거
installation ; 설치, 설비
resemble ; 닮다
phillips screw ; 필립스 나사
leverage ; 지레장치. 지레의 작용
tighten ; 죄다, 죄이다
loosen ; 풀다, 늦추다. 느슨해지다.
notch ; (V형 또는 톱니모양)의 새긴 금,
disadvantage ; 불리, 단점
restricted space ; 제한된 공간
adjustable ; 조절할 수 있는.
jaw ; 턱
odd size ; 이상한 치수(크기)
detachable ; 분리할 수 있는
torque ; 토크, 비트는 힘
damage ; 손해, 손상
register ; 등록, 등록부
plier ; 집게, 펜치
pivot ; 주축, 중심점, 요점
grip ; 꽉잡다, 손잡이
hex ; 육(六)의 뜻의 결합어
roughen ; 거칠게 하다[되다]
claw hammer ; 못 빼는 망치
ball pin hammer ; 볼핀햄머
chisel ; 끌, 끌로 파다.
goggle ; (용접 등의) 보호 안경.
chip ; 조각, 토막
workpiece ; 제조공정에 있는 소재
mushroom ; 버섯
file ; 줄
classify ; 분류하다.
triangular ; 삼각형의
rivet ; 리벳, 대갈 못
align ; 일직선으로 하다.
hacksaw ; 쇠톱
replaceable ; 대치할 수 있는

명언 / 격언

Slow and steady wins the race.

— *Robert Loyd*

2

Chapter

Measurements and Measuring Tools

(측정 및 측정공구)

2.1 RULER

The common Ruler or steel scale is marked off in inches and fractions of an inch. Sometimes these markings are as small as 1/64 inch. Other rulers or steel scales are marked with both inches and centimeters, or millimeters(Fig. 2.1). The metric ruler has only metric markings.

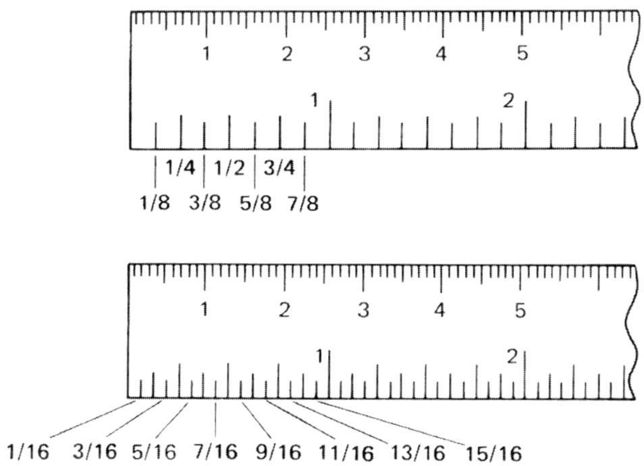

Fig. 2.1 Ruler, or steel scale, marked in both inches and centimeters

Chapter 2

2.2 FEELER GAGES

Feeler gages are strips or blades of hardened steel or other metal, ground or rolled with extreme accuracy to the proper thickness. They are generally supplied in sets (Fig. 2.2), with each blade being marked with its thickness in thousandths of an inch, thousandths of a millimeter, or both. For example, in Fig. 2.2, the gage marked 0.003 inch is also marked in its metric equivalent of 0.08 millimeter.

A set of feeler gages Fig. 2.2

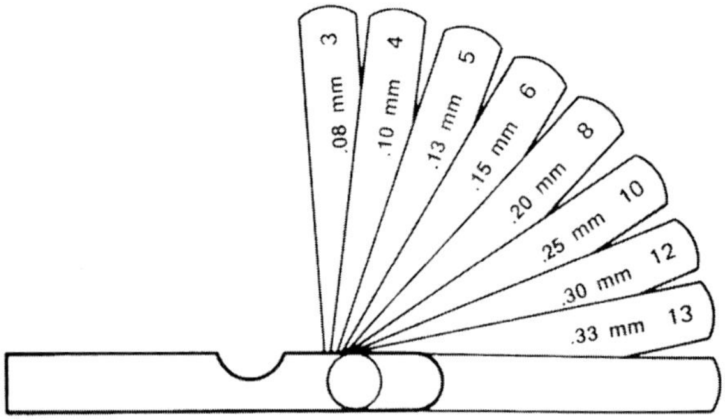

2.3 WIRE FEELER GAGES

Wire feeler gages (Fig. 2.3) are similar to the flat feeler gages, except that they are made of carefully calibrated steel wire of the proper thickness. They are useful in checking sparkplug gaps and similar dimensions. Metric wire feeler gages are also available.

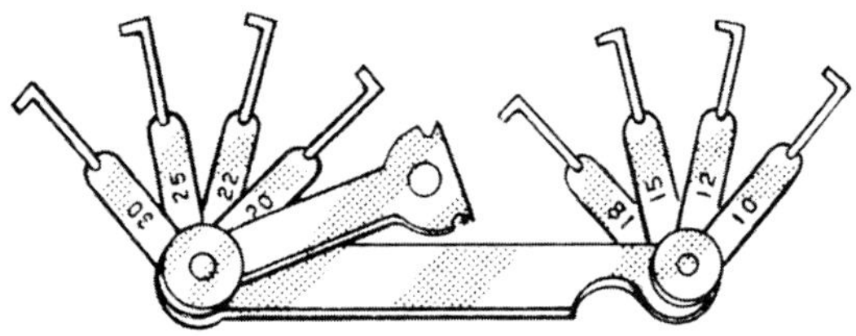

Fig. 2.3 A set of wire feeler gages

2.4 CALIPERS

Calipers (Fig. 2.4) can be used with a scale to make fairly accurate measurements of many parts. The **outside caliper** can be used for measuring the diameter of a shaft. First, it is adjusted by turning the thumb nut until the legs will just slip over the shaft with a slight drag. Then, the distance between the two legs can be measured with a scale.

In a similar manner, the **inside caliper** can be used to measure the diameter of a hole. It is adjusted until the legs can be moved in the hole with a slight drag. Then, the distance between the legs is checked with a scale.

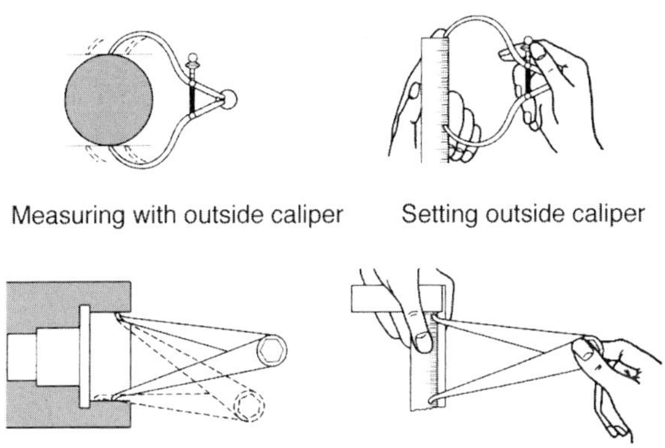

Measuring with outside caliper · Setting outside caliper

Measuring with inside caliper · Setting inside caliper

Fig. 2.4 Calipers

2.5 MICROMETER

Micrometer is one of the most familiar precision measuring devices used in the workshop and the inspection room. Fig. 2.5 shows the construction of a typical micrometer and names and describes the more important parts.

The operation of a micrometer depends upon the principle that the distance moved by a nut along a screw is proportional to the number of revolutions made by the nut.

A micrometer Fig. 2.5

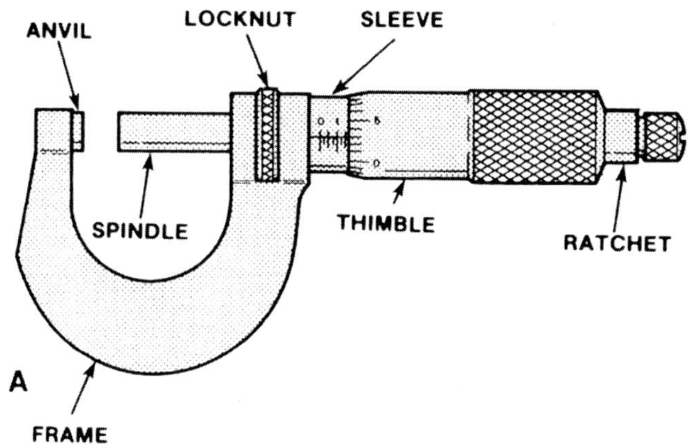

Fig. 2.6 shows the scales for a metric micrometer with a screw which has a lead of 0.5mm. The illustrated reading is as follows:

	mm
9 whole millimeters	9.00
1 half millimetre	0.50
48 hundredths of a mm	0.48
Total reading	9.98

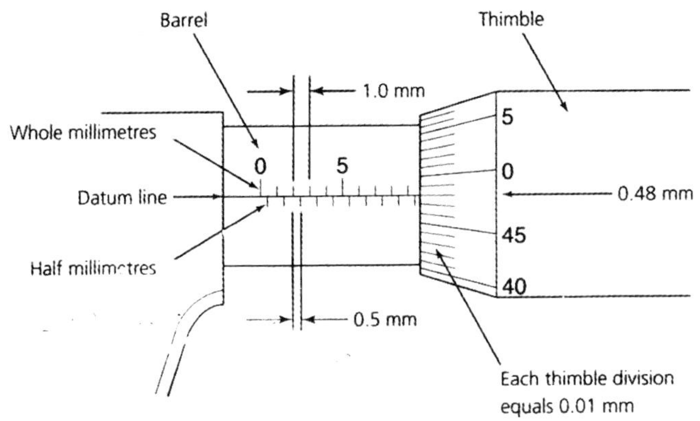

Fig. 2.6 Micrometer scales (metric)

2.6 VERNIER CALIPER

Fig. 2.7(a) shows a vernier caliper. Unlike the micrometer, the same instrument can take inside and outside measurements, as shown in Fig. 2.7(b). The vernier caliper reads from zero to the full length of its beam scale, whereas the micrometer only reads over a range of 25mm.

Fig. 2.7 Vernier caliper: (a) design; (b)use of internal and external jaws; (c) the scale

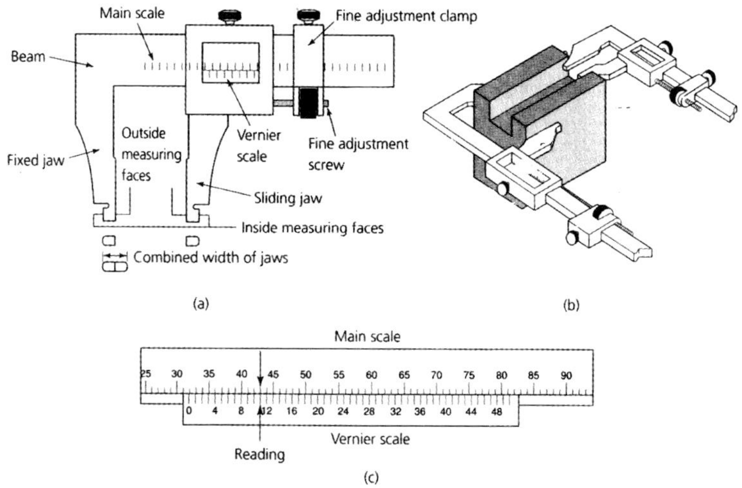

Unfortunately it is difficult to use a vernier calipers as accurately as a micrometer for the following reasons:

- It is difficult to obtain a correct feel due its size and weight.
- The scales are difficult to read even with the aid of a magnifying glass.

Fig. 2.7(c) shows a typical 50-division vernier scale as used on a metric instrument. The zero mark of the vernier scale is just beyond the 32 on the main scale, so the measurement is slightly in excess of 32mm. The first pair of vernier and main scale graduations to come into line with each other is at 11 on the vernier scale. For the scales shown, since each vernier division is 0.02mm, the total reading is as follows:

	mm
32 whole millimeters	32.00
11 vernier divisions	0.22
Total reading	32.22

Chapter 2

VOCABULARY

measurement ; 측정
measuring tools ; 측정공구
ruler ; 자(=scale)
fraction ; 분수, 단편
grind ; 갈다, 갈아서 닳게 하다
roll ; 굴리다, 금속을 압연하다.
accuracy ; 정확성, 정밀도
thickness ; 두께, 굵기
supply ; 공급하다
equivalent ; 동등한, 상당(대응)하는
calibrate ; 눈금을 정하다.
available ; 이용할 수 있는
caliper ; 캘리퍼스
micrometer ; 측미계(測微計)
vernier ; 버니어, 부척(副尺)
precision ; 정확, 정밀
inspection ; 검사, 점검
principle ; 원리, 원칙
proportional ; 비례하는
range ; 범위
magnify ; 확대하다.
division ; 분할, 구분
excess ; 초과, 과잉
graduation ; 대학졸업, 눈금

명언 / 격언

Give me a long enough lever and a place to stand, and I will move the Earth.

— *Archimedes*

3 Chapter Technical Drawing

(공업제도)

3.1 TYPES OF LINES

Twelve types of lines that are used on technical drawings are illustrated and described in Fig. 3.1.

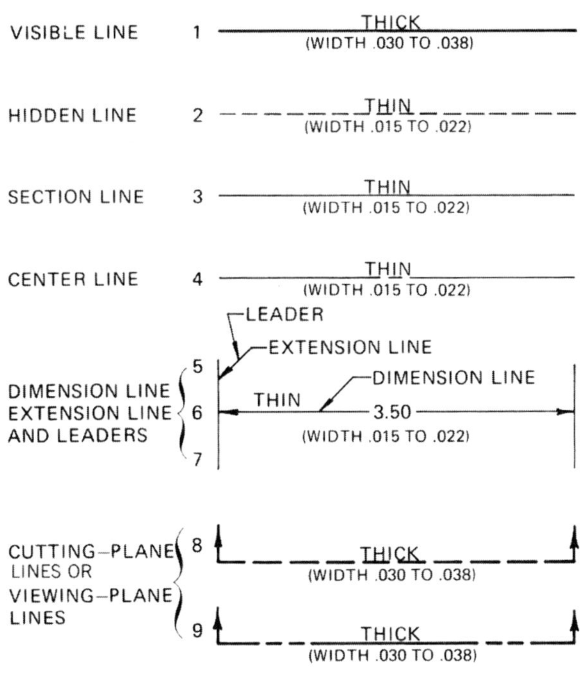

Fig. 3.1 Width and type of lines (KS)

Chapter 3

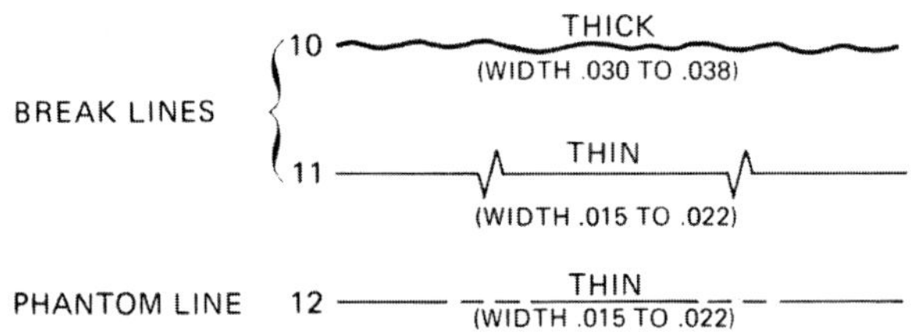

The **visible outline** in Fig. 3.1 serves to show you what the object looks like. It is a thick line. The width of the line makes the outline of the object seem to stand out on the drawing. Hidden details are shown by lines made up of short dashes.

These 12 lines are so important that they are also called the alphabet of lines.

3.2 APPLICATION OF TYPES OF LINES

In Fig. 3.2 each of the 12 types of lines shown in Fig. 3.1 is illustrated in the drawing of a machine part. It is important that you study each line separately. The best way for you to learn the use of these lines would be to make a sketch or tracing of Fig. 3.2 and label each line.

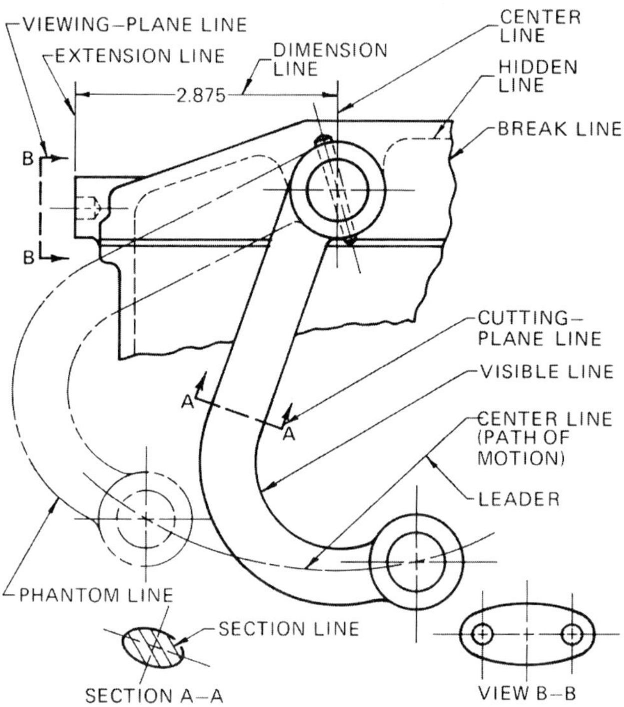

Fig. 3.2 Application of types of lines

3.3 FIRST-ANGLE AND THIRD-ANGLE

The terms **first-angle** and **third-angle** are derived from the mathematician' s convention of annotating the four right angles which make up the 360°of a circle, the first angle being 0° to 90° and the third angle from 180°to 270°(Fig. 3.3).

First-angle and third-angle Fig. 3.3

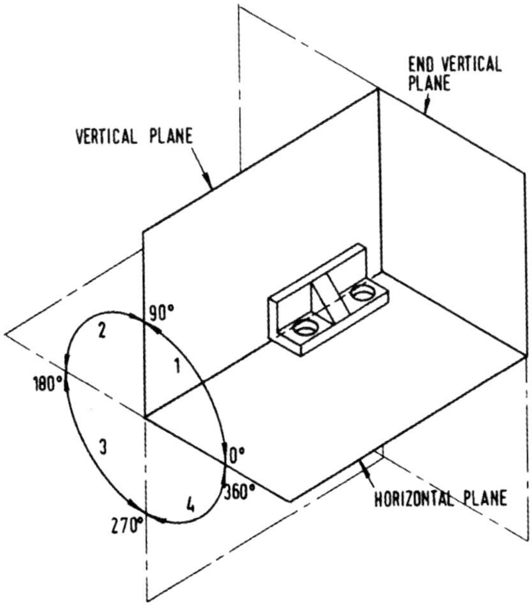

3.4 FIRST-ANGLE PROJECTION

First-angle projection is almost universally followed in Europe and was used in the United states until around 1890. In principle, projection in the first-angle means that when the image planes are revolved into the plane of the paper the top view will appear **below** the front view and the left-side view will appear to the **right** of the front view (Fig. 3.4).

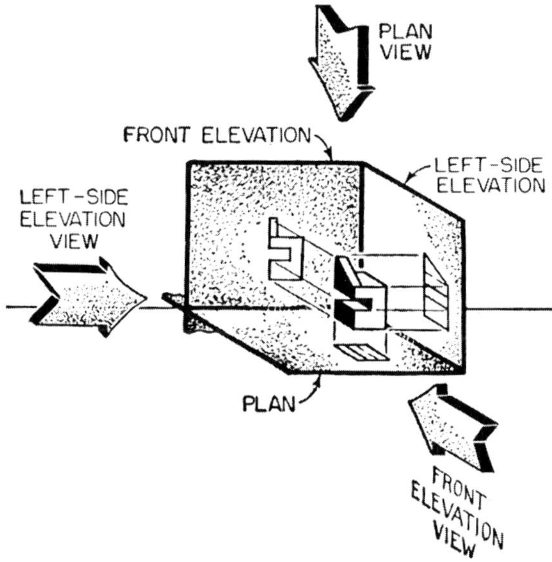

3.4 First-angle projection

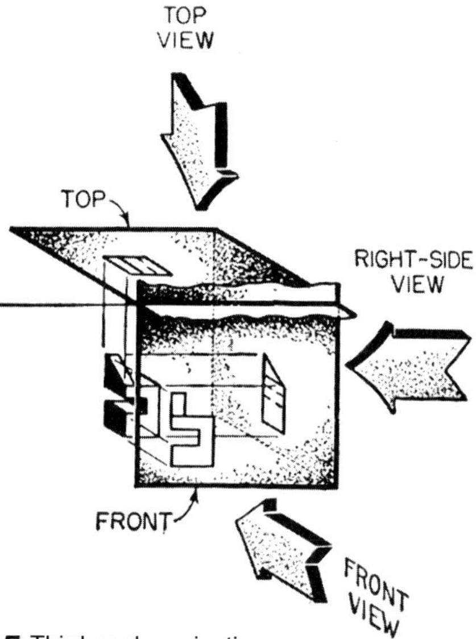

3.5 Third-angle projection

3.5 THIRD-ANGLE PROJECTION

Third-angle projection is the system of projection used in this chapter and throughout the text. In essence, third-angle projection (Fig. 3.5) means that the plane of projection is always placed between the observer and the object (that is, the object is viewed by looking **through** the projection plane).

3.6 SECTIONING

To produce a sectional view, a **cutting plane** is assumed to be passed through the part for the design, as shown in Fig. 3.6(a).

Then, at(b) the cutting plane is removed and the two halves drawn apart, exposing the interior construction. In this case, the direction of sight is toward the left half, as shown, and for purposes of the section the right half is mentally discarded. The sectional view will be in the position of a right-side view.

Fig. 3.6 A section

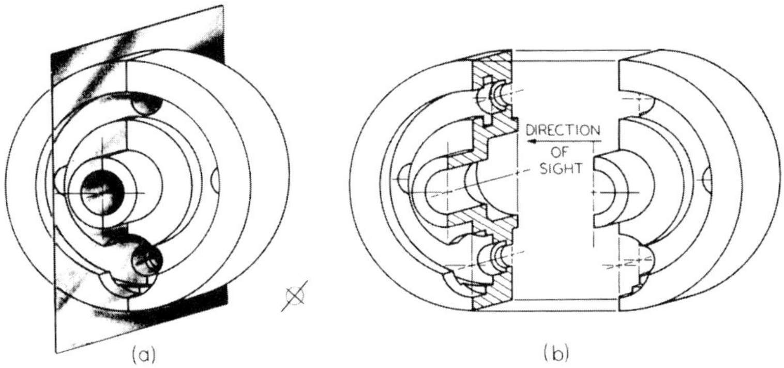

3.6.1 FULL SECTIONS

Sectional views as shown in Fig. 3.7 are known as full sections; that is, the object is represented as being cut entirely through. Arrows on section plane show direction in which sectioned view is taken.

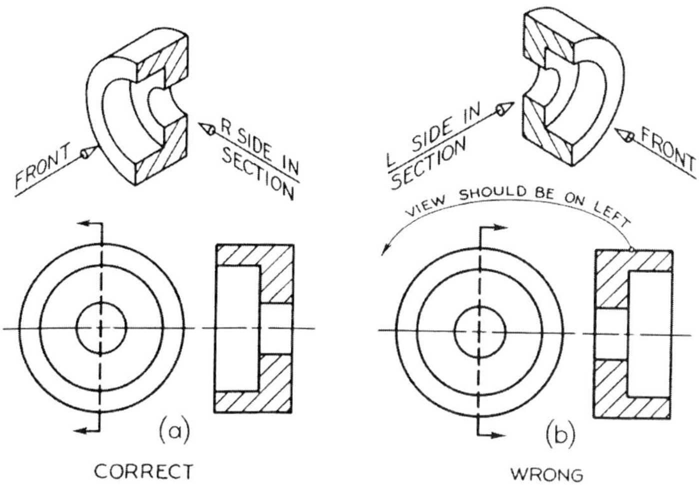

Fig. 3.7 The full section

3.6.2 OFFSET SECTIONS

In sectioning through irregular objects, it is often desirable to show several features that do not lie in a straight line, by "offsetting" or bending the cutting plane. Such a section is called an offset section (Fig. 3.8).

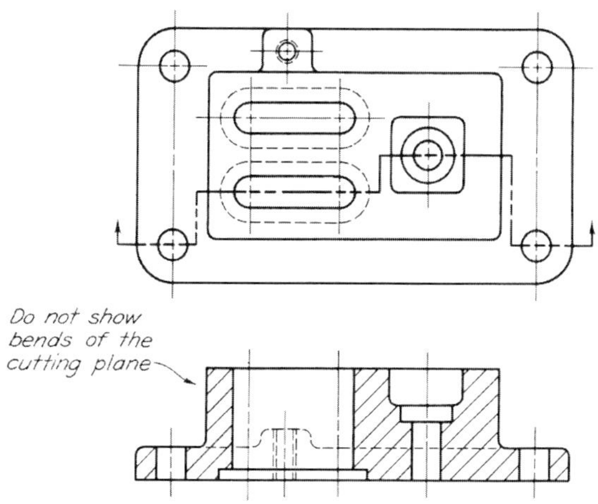

Fig. 3.8 Offset section

3.6.3 HALF SECTIONS

At times it is necessary to represent an object as if were cut only halfway across(Fig. 3.9). The remaining portion would be shown as a regular view. Such a sectional view is called an half section (Fig. 3.10). This type view is used particularly in describing a symmetrical object in which both halves of a full section would be alike. In general, hidden lines should be omitted from both halves of a half section. However, they may be used in the unsectioned half if necessary for dimensioning.

Cutting plane Fig. 3.9

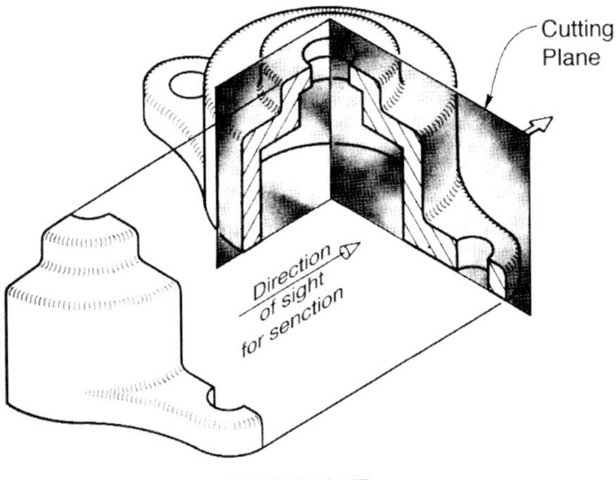

CUTTING PLANE

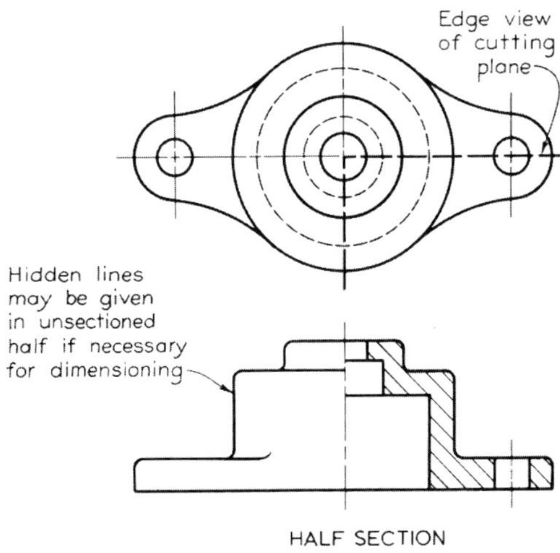

Fig. **3.10** The half section

3.6.4 REVOLVED SECTIONS

The revolved section is used when it is difficult to represent the true shape of a part at a certain point. This represents the shape of the object, at a particular point, upon the regular view. (See Fig. 3.11).

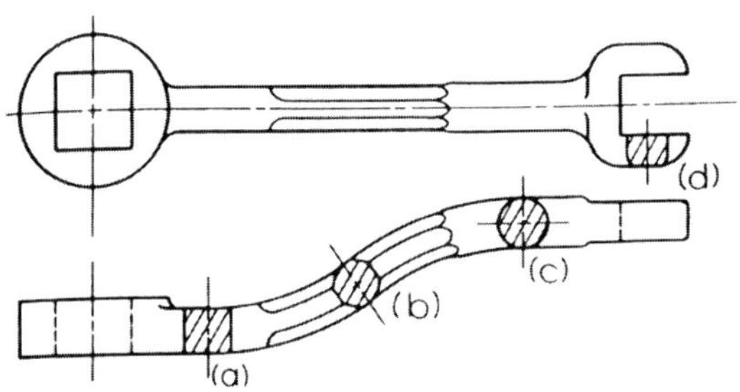

Fig. **3.11** Revolved sections

The true shape of a revolved section should be retained after the revolution of the cutting plane, regardless of the direction of the lines in the view (Fig. 3.12).

A common error in drawing revolved. *Fig.* **3.12**

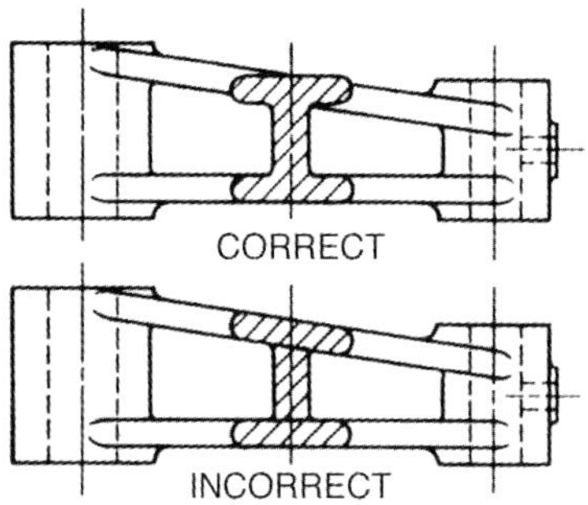

3.6.5 REMOVED SECTIONS

A front view of the crane hook is shown in Fig. 3.13. The crane hook is typical of objects which are not symmetrical and which change shape gradually from one end to the other. In many instances there is not room enough on the front view to show as many revolved sections as would be necessary for complete representation. The **revolved section** must be removed from the view itself and placed elsewhere on the sheet. Hence, we call the name **removed section.**

The removed section *Fig.* **3.13**

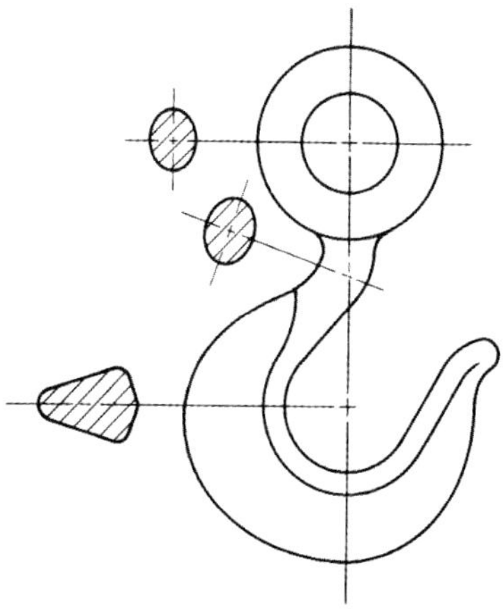

3.6.6 BROKEN-OUT SECTIONS

Broken-out sections are used when it is not desirable to make a complete sectional view. In these instances the details of the construction may be shown by a section drawn on the object at the position where it is desirable, as is shown in Fig. 3.14.

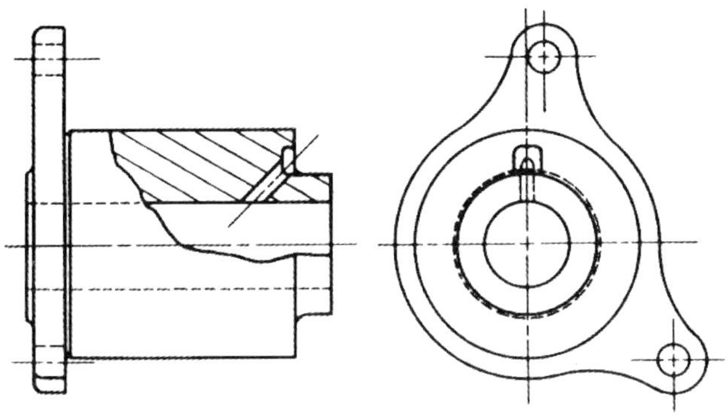

3.14 Broken-out section

3.6.7 RIBS IN SECTION

When a cutting plane passes longitudinally (lengthwise) through the center of rib, web, gear tooth, or similar flat element, the element should not be sectioned (Fig. 3.15). This will avoid a false impression of thickness.

Ribs in section Fig. 3.15

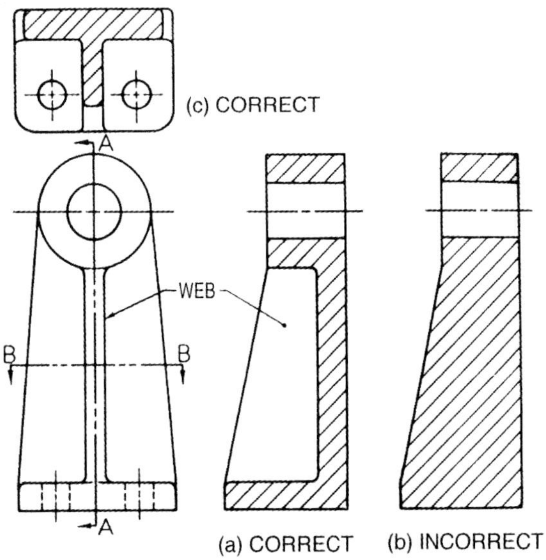

3.6.8 ALIGNED SECTIONS

Spokes, arms, and holes frequently appear in odd numbers and will give an unsymmetrical and misleading section if the rules of true projection are followed. An example of this is shown in Fig. 3.16. The wheel has been sectioned on the vertical center line of the front view, and a true sectional view is shown at the right.

Spokes should never be sectioned, whether they occur in the line of the cutting plane or whether they are imagined to be revolved into it. When elements such as spokes, holes, and arms are revolved into the line of the cutting plane, they are said to be **aligned.**

The three arrows show how a spoke, a keyway, and a hole in the front view should be revolved into the line of the cutting plane in order to make a conventional right-side sectional view

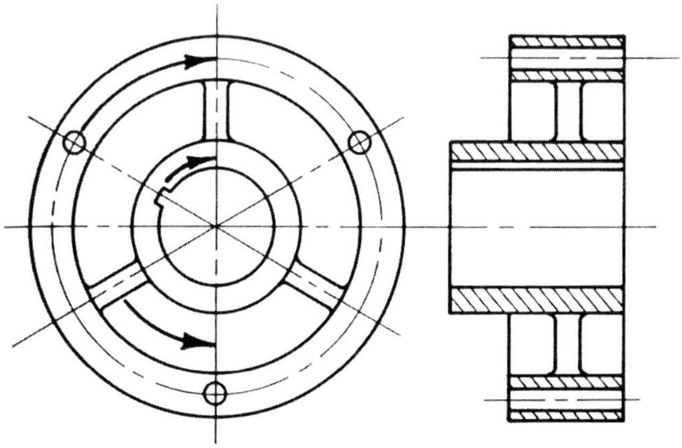

3.16 Aligned section

shown in Fig. 3.16.

3.6.9 THIN SECTIONS

Sections too thin (less than 1/8 in. thick) for crosshatching may be shown solid. When two or more adjacent parts are shown

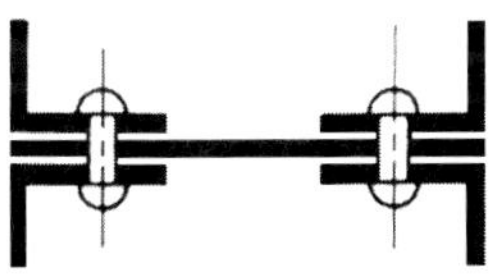

3.17 Thin section

solid, a space should be left between the parts (Fig. 3.17).

3.6.10 CONVENTIONAL BREAKS

Five different conventional breaks are shown in Fig. 3.18(a) through (e). A break in a solid round bar is shown at (a). Breaks in a round tubular piece (pipes) are shown at (b) and (c). A break in a rectangular piece of metal is shown at (d), and a break in a rectangular piece of wood is shown at (e).

Conventional breaks *Fig.* **3.18**

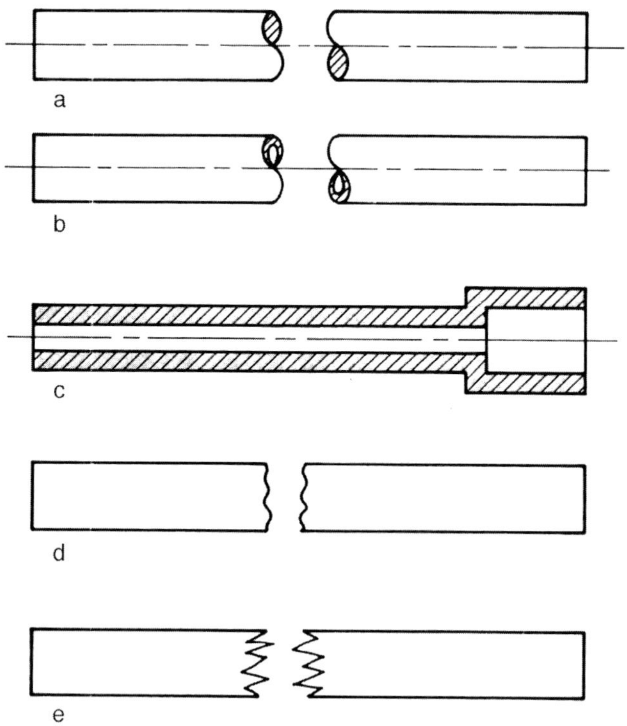

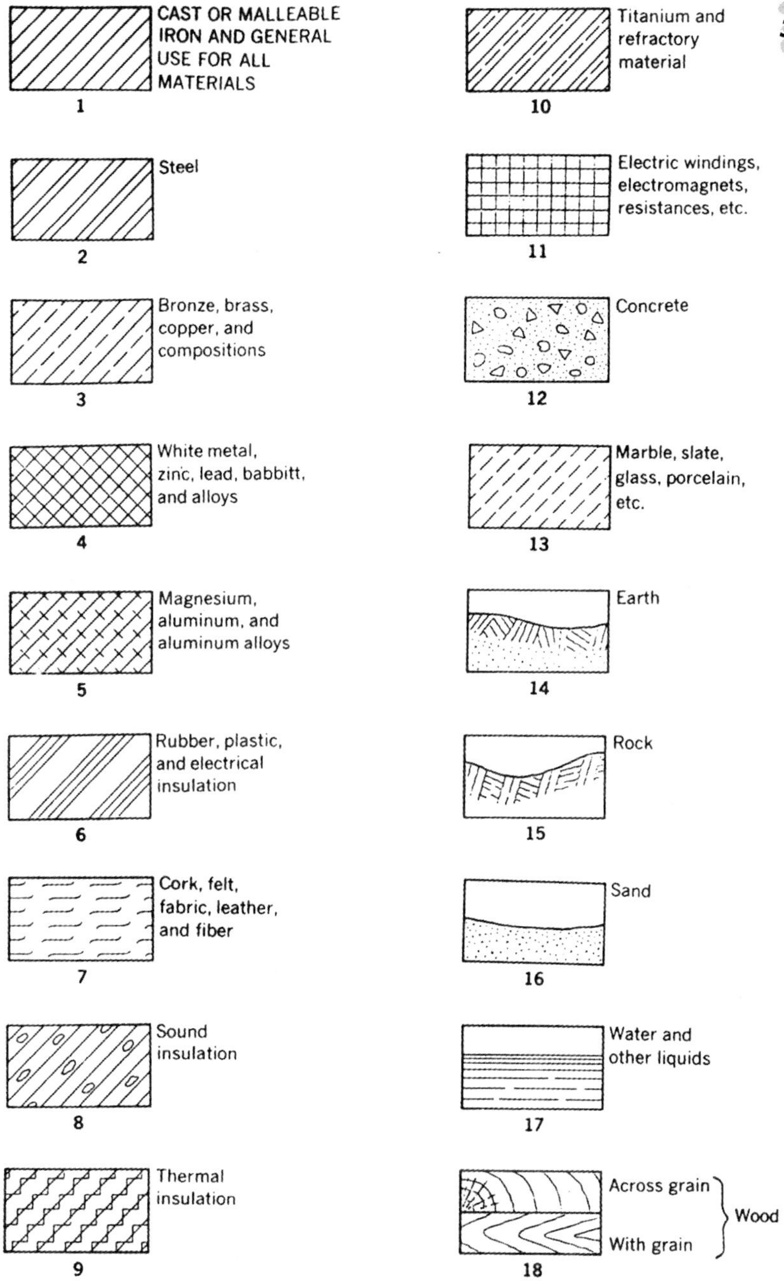

Fig. 3.19 Symbols for section lining

3.7 SYMBOLS FOR SECTION LINING

Section lining may be used to show materials of which an object is made as well as its interior construction. The section lining for cast iron in Fig. 3.19 is called the universal section lining and is frequently used on detail drawing of individual parts whether they are made of cast iron or not.

However, on many architectural, electrical, structural, and other types of drawings it is advisable to use the American Standard symbols for section lining. These are shown in Fig. 3.19.

3.8 TOLERANCE

The difference between the two limits of a dimension of a part is called tolerance on that dimension. For example, a dimension given as 1.625 ±.002 means that it may be (on the manufactured part) 1.627″ or 1.623″, or anywhere between these limit dimensions. The tolerance is .004″.

Tolerances must be placed on the dimensions of a drawing to limit the permissible variations in size because it is impossible to manufacture a part exactly to a given dimension. Although small tolerances give higher quality work and a better operating mechanism, the cost of manufacture increases rapidly as the tolerances are reduced , as indicated by the typical curve of Fig. 3.20.

It is therefore important that the tolerances be specified at the largest values that the operating or functional considerations permit.

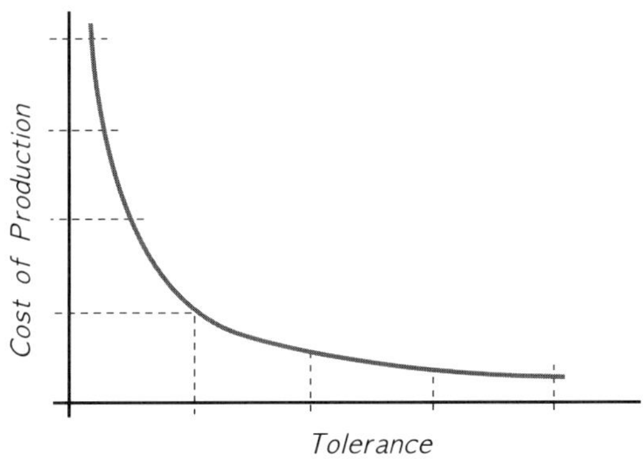

Fig. 3.20 Relation between size of tolerance and cost of production

3.9 TOLERANCE DISPLAY SYSTEM

Tolerance may be displayed by three system. The three methods are,

(a) Unilateral Tolerance

(b) Bilateral Tolerance

(c) Limit Dimensioning

Unilateral Tolerance. In the unilateral method, the tolerance is shown in one direction only away from the design size.

Examples are:

$$1.500 \pm {}^{.002}_{.000} \text{ or } .675 \pm {}^{.000}_{.003}$$

Bilateral Tolerance. In the bilateral method, the tolerance is shown in both directions away from the design size. Since the tolerance is shown in both directions away from the design size it is possible to have two methods for displaying bilateral tolerances.

These are,

(1) Unequal bilateral

(2) Equal bilateral

Examples of unequal bilateral tolerances are:

$$1.500\pm{}^{.002}_{.005} \text{ or } .675\pm{}^{.004}_{.001}$$

Examples of equal bilateral tolerances are:

$$1500\pm.002 \text{ or } .100\pm.001$$

Limit Dimensioning. In the limit dimensioning method the maximum and minimum allowable sizes are specified and the difference between them is the tolerance.

Examples of limit dimensioning are:

$$\begin{matrix}1.502\\1.498\end{matrix} \text{ or } \begin{matrix}.376\\.375\end{matrix}$$

3.10 CLASSES OF FITS

Fits on machine parts are of three general types: the clearance fit, the intereference fit, and the transition fit (Fig. 3.21).

Clearance fit. In the clearance fit, the internal member is smaller than the external member, and the mating parts move rather freely. The space between the parts allowing this freedom is termed the "allowance" and is indicated as positive.

Interference fit. In the interference fit there is interference of metal because the internal member is larger than the external member, and the allowance is negative.

Transition fit. A transition fit is one in which the parts are made to allow either a clearance or an interference fit—a maximum external member and a minimum internal member will give a clearance fit; a minimum external member and a maximum internal member will give an interference fit.

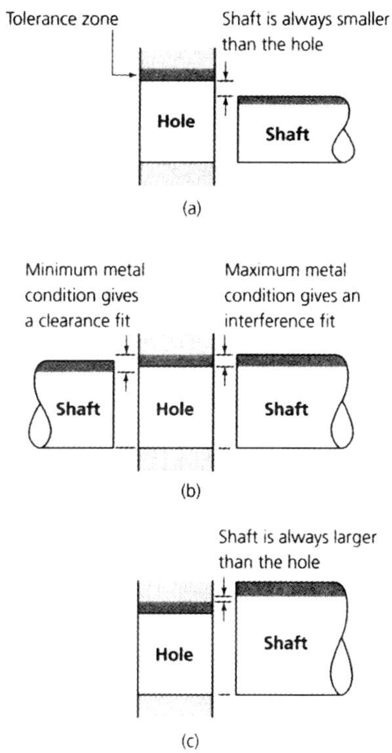

Fig. **3.21** Classes of fit:
(a) Clearance:
(b) Transition:
(c) Intereference:

Chapter 3

VOCABULARY

illustrate ; 설명하다. 삽화를 넣다.
describe ; 묘사하다, 그리다
visible line ; 외형선
hidden line ; 숨은선
section line ; 단면선
center line ; 중심선
dimension line ; 치수선
extension line ; 연장선
leader ; 인출선, 지시선
break line ; 파단선
phantom line ; 가상선
detail ; 세부, 상세도
dash ; 돌진, 대시(-)
label ; 꼬리표, 딱지, 분류표시
first-angle projection ; 제1각투영법
revolve ; 회전시키다.
front view ; 정면도
third-angle projection ; 제3각투영법
full section ; 전단면도(全斷面圖)
represent ; 다시 나타내다
offset section ; 편위 단면도
irregular ; 불규칙적인, 변칙적인
half-section ; 반단면도
clearance fit ; 헐거운 끼워맞춤
interference fit ; 억지 끼워맞춤
transition fit ; 중간 끼워맞춤
interchangeable ; 교환할 수 있는
revolved section ; 회전 단면도
removed section ; 이동 단면도
hook ; 갈고리, 걸쇠
gradually ; 점차로, 차차
instance ; 실례, 사례
partial section ; 부분 단면도
longitudinally ; 세로의
aligned section ; 정열 단면도
spoke ; 차바퀴의 살
conventional ; 관습적인
tubular ; 관모양의
architectural ; 건축상의
advisable ; 권할만한
manufacture ; 제작, 제조공업
permissible ; 허용할 수 있는
variation ; 변화, 변동, 변화량
tolerance ; 공차, 허용오차
specify ; 설계명세서에 기입하다.
display ; 진열하다. 전시하다.
unilateral tolerance ; 편측공차
bilateral tolerance ; 양측공차
Limit dimensioning ; 한계치수 방식
portion ; 한조각, 일부
particularly ; 특별히
symmetrical ; 대칭적인, 균형잡힌
omit ; 빠뜨리다, 생략하다.

Chapter 4

Choosing a Material

(재료선택)

4.1 INTRODUCTION

When designing a mechanism, different materials are often chosen from which to manufacture the individual components. The purpose of this section is to consider the factors that lead to the choice of material for a particular component.

The body of the electric plug in Fig. 4.1 is manufactured from a plastics material. One reason for the choice of this material is that it is a good **electrical insulator.** The pins of the plug must be good **electrical conductors.** Copper is a good conductor of electricity but it is **soft** and **weak**, so brass, which is **harder** and **stronger** than copper and almost as good at conducting electricity, is used for the pins.

Chapter 4

Three-pin electric plug Fig. 4.1

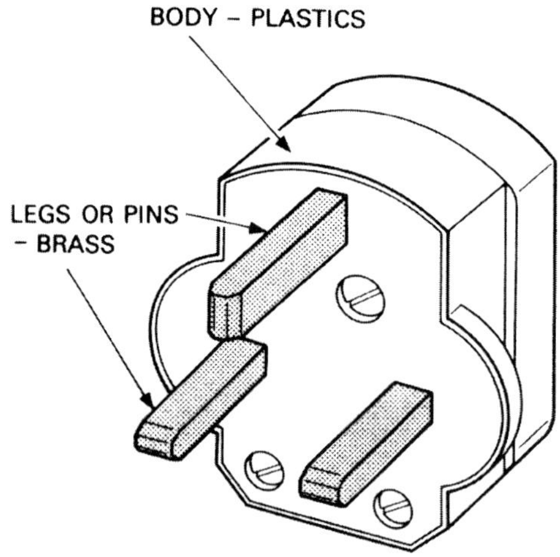

The above factors, which describe the behaviour of materials under certain condition-hardness, strength, electrical insulation, electrical conductivity, etc.—are described as the **properties** of a material. It is the properties of a material that make it suitable or unsuitable for component manufacture.

Properties such as electrical conductivity or electrical insulation, and heat conductivity or insulation are obvious to most people, but engineers are usually concerned with less obvious mechanical properties.

4.2 MECHANICAL PROPERTIES

Some of the mechanical properties of a material are as follows:

(1) Strength

(2) Toughness

(3) Hardness

(4) Ductility

(5) Malleability

Strength. The strength of a material is a measure of how well a material can withstand a steady load or force without breaking. The loads on the tow rope in Fig. 4.2 are tending to pull the rope apart.

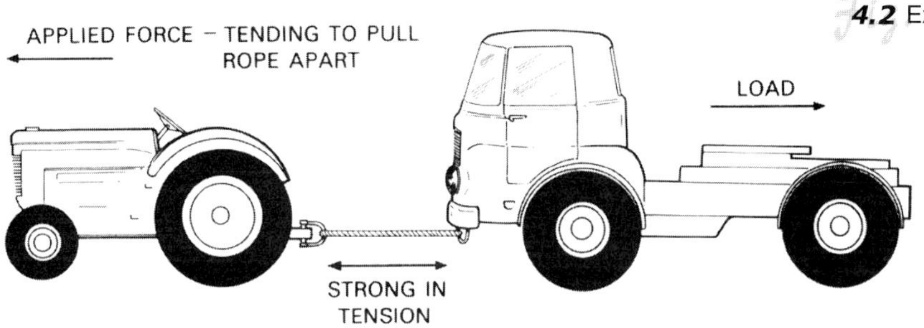

Fig. **4.2** Example of tensile loads

Such stretching loads need good properties of **tensile** strength in the material. Loads that tend to compress a component (Fig. 4.3) need a material that has good **compressive** strength, and the twisting forces (Fig. 4.4) good shear strength. The tensile, compressive and **shear** strength, etc., of a material generally differ, for example, cast iron has good compressive strength properties but poor shear-strength properties.

Compressive strength *Fig.* **4.3**

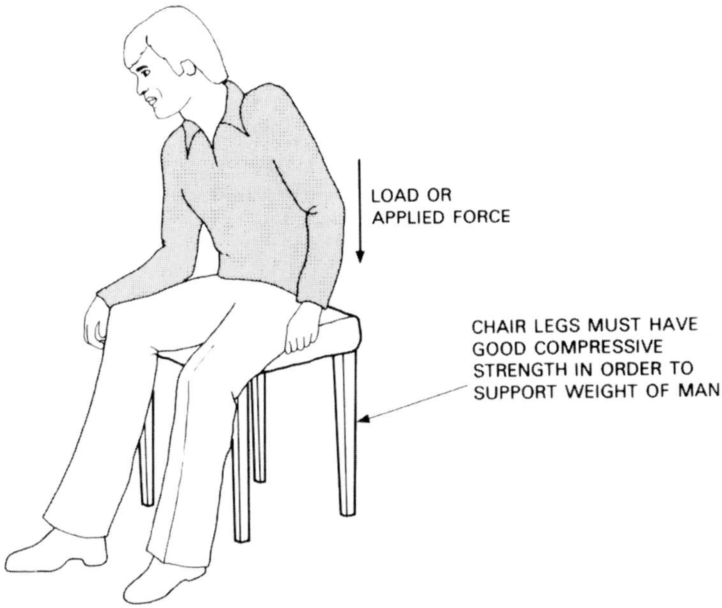

Twisting forces *Fig.* **4.4**

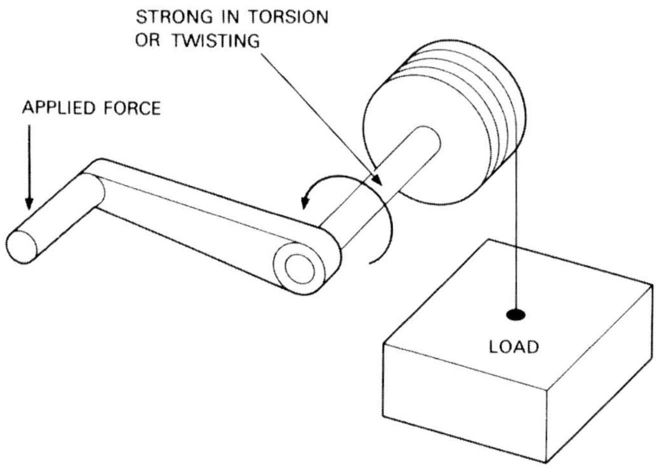

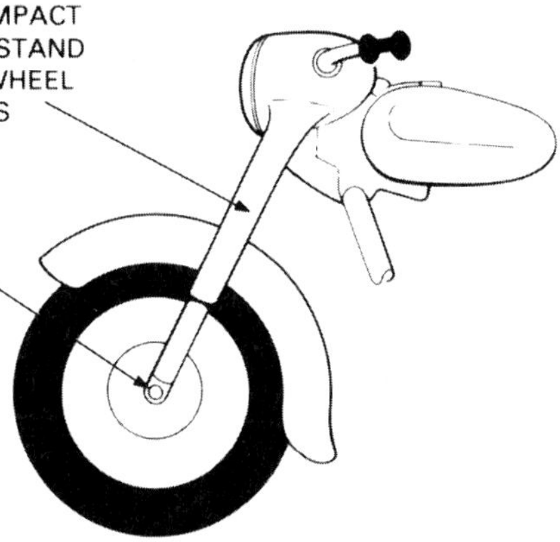

Fig. **4.5** Impact strength or toughness

Toughness. Strength properties take into account the behaviour of material when it is subjected to a steady load or force. The ability of a material to withstand sudden or shock loads is the property of **impact strength** or toughness.

The materials used in the construction of the motorcycle front forks (Fig. 4.5) must have good toughness properties to withstand the impacts shown. If a material has poor toughness properties it is said to be brittle.

Hardness. If the drill in Fig. 4.6 is to cut the metal and produce the hole it must be **harder** than material it is cutting. A material that has poor harness properties is a **soft** material. Where components rotate or slide relative to each other, the softer material will wear more quickly than the hard material.

Hardness Fig. 4.6

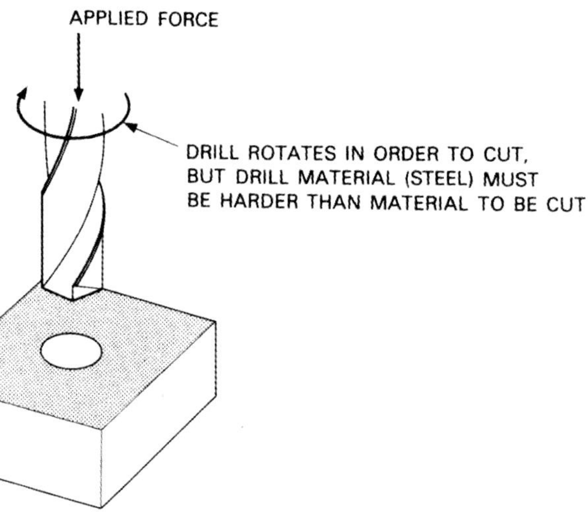

Ductility: principle of wire drawing Fig. 4-7

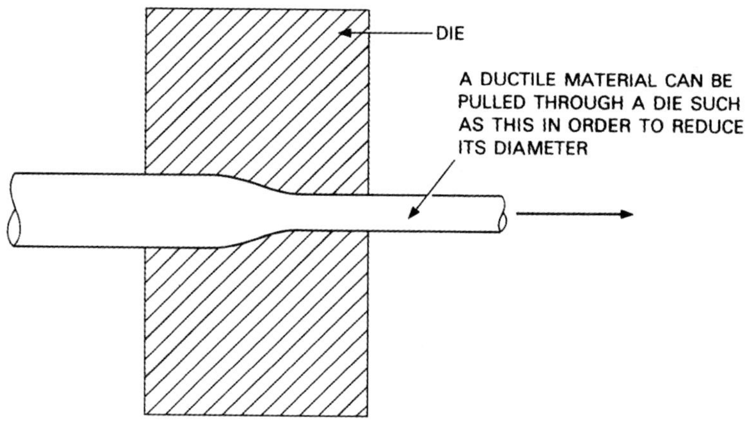

Ductility. If a material can be pulled into a longer but thinner section (Fig. 4.7), it is said to have good ductility properties.

Malleability. Fig. 4.8 shows a material being hammered into shape; good malleable properties are a material to be beaten into a flat from.

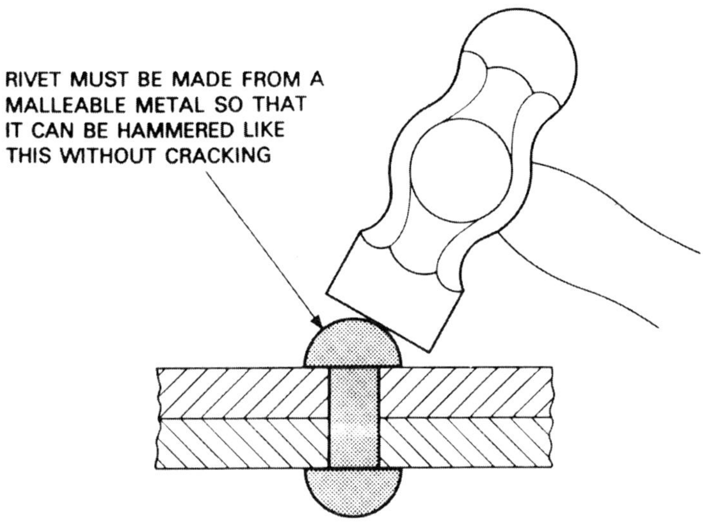

Fig 4.8 Malleability

4.3 ENGINEERING MATERIALS

The many materials used in engineering can be classified as follows.

(1) Ferrous Metals.

(2) Non-ferrous Metals.

(3) Non-metallic Materials.

4.4 FERROUS METALS

Ferrous metals all contain the metal **iron.** Pure iron is too soft and weak for most engineering application, so it is "alloyed" (mixed) with carbon to produce the most useful and common engineering metals-steel and cast iron.

Small variations in the amount of carbon mixed with the iron will make great differences to the properties of the metal. The following examples show some of the more common **iron-carbon** alloys and their uses.

(1) **Cast iron** : 3 to 3.5 per cent carbon mixed with iron.

Cast iron is easily **cast** into complicated shapes (Fig. 4.9); it has good compressive strength properties but is brittle.

Fig. **4.9** Typical sand-cast component

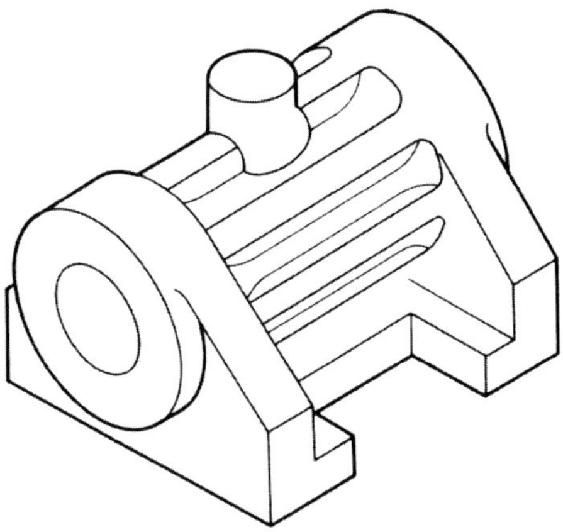

(2) **High-carbon steel** : 0.8 to 1.5 per cent carbon.

This material is the hardest and strongest of the steels ; its hardness can be increased by heat treatment ; knives, files and some drills are manufactured from high-carbon steel.

(3) **Medium-carbon steel** : 0.3 to 0.8 per cent carbon.

This is used if a component needs a steel with good toughness properties; leaf springs and axles are typical applications.

(4) **Mild steel** : 0.1 to 0.3 per cent carbon.

The cheapest of the carbon steels; it can be shaped easily by bending and pressing. Applications include car bodies, thin wire and girders ; where hardness or strength properties are not too important, mind steel is a good general-purpose steel.

(5) **Alloy steels**

These are carbon steels to which small quantities of other metals have been added to produce specific properties; an example is stainless steel, where chromium is added to give properties of corrosion resistance.

4.5 NON-FERROUS METALS

Non-ferrous metals all do not contain the metal **iron.** Some of the important "pure" non-ferrous metals are listed in table 2.1

Table 2.1

Pure non-ferrous metals	Typical properties
Aluminium	Light in weight: week and soft
Copper	Good conductor of electricity and heat; soft and weak: corrosion resistant
Zinc	Brittle; easy to cast
Tin	Soft; low melting point: corrosion resistant
Lead	Heavy in weight: soft; low melting point

These pure metals have limited use in engineering, but when two or more are alloyed it is possible to remove or reduce undesirable properties while at the same time retaining desirable ones. For example

(1) **Brass** : an alloy of copper and zinc; a typical application is shown in Fig. 4.10.

Fig. 4.10 Domestic water-tap - brass casting; 60% copper, 40% zinc.

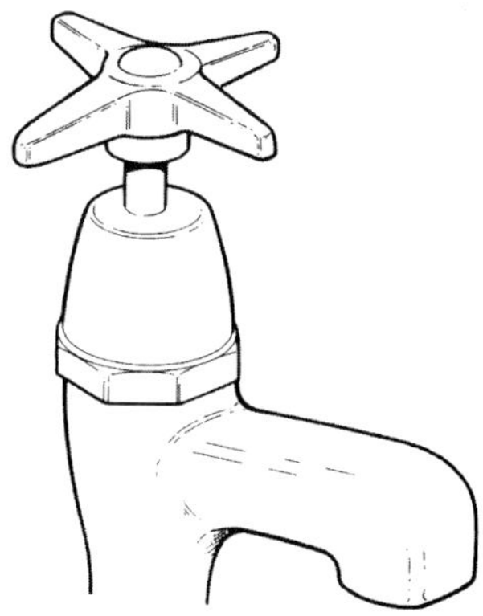

(2) **Bronze** : alloys of copper and non-ferrous metals other than zinc are usually grouped together under the title of bronze; examples are shown in Fig. 4.11, 4.12 and 4.13.

Fig. 4.11 Marine pump body

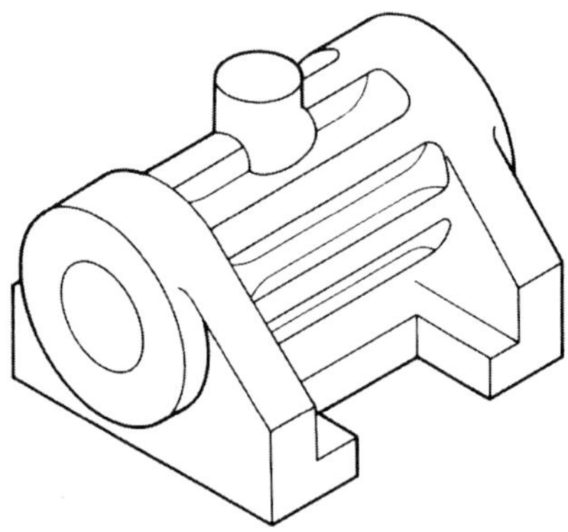

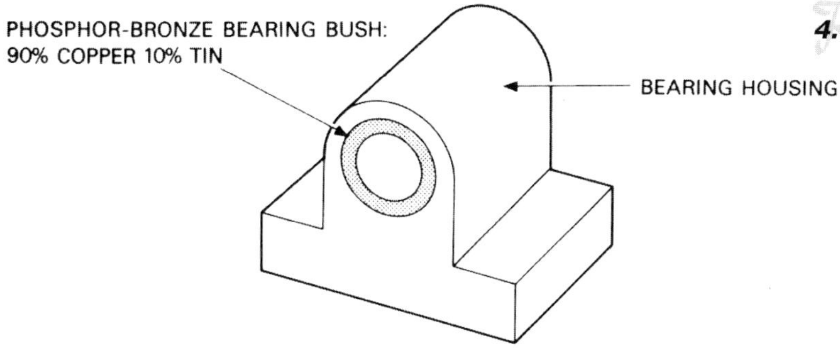

4.12 Bearing housing

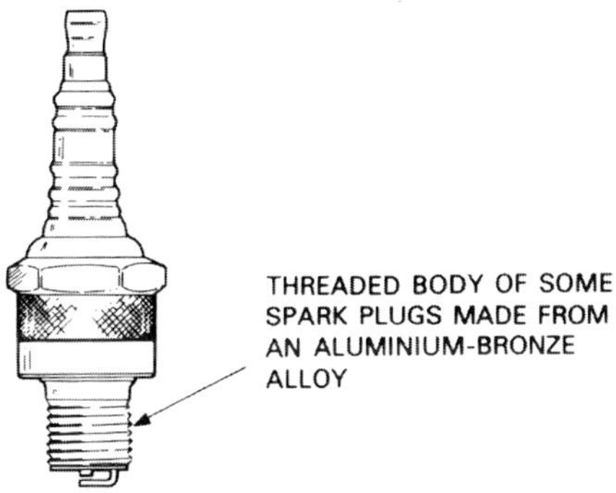

4.13 Spark plug

(3) **Aluminium alloys** : the lightness of aluminium makes the material suitable for aircraft construction, but it is first alloyed with other materials to improve its strength and hardness; examples are shown in Fig. 4.14 and 4.15.

Lightweight *Fig.* **4.14** motorcycle engine

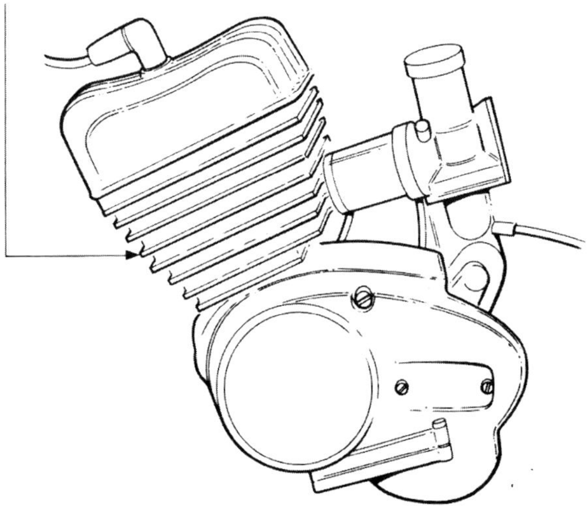

Aircraft undercarriage *Fig.* **4.15**

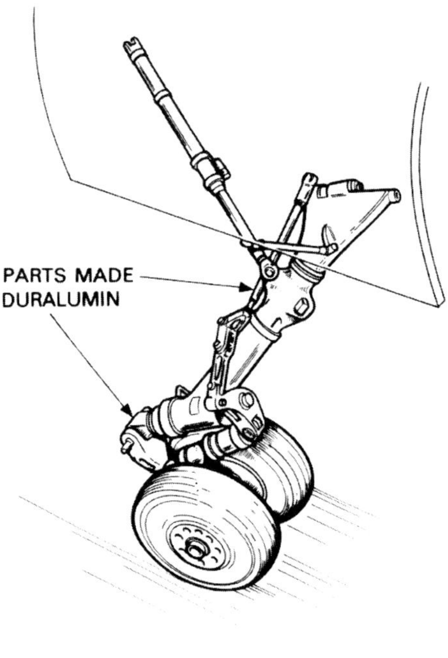

(4) Zinc-based alloys : these materials are used in low-melting-point diecasting applications, such as the example shown in Fig. 4.16.

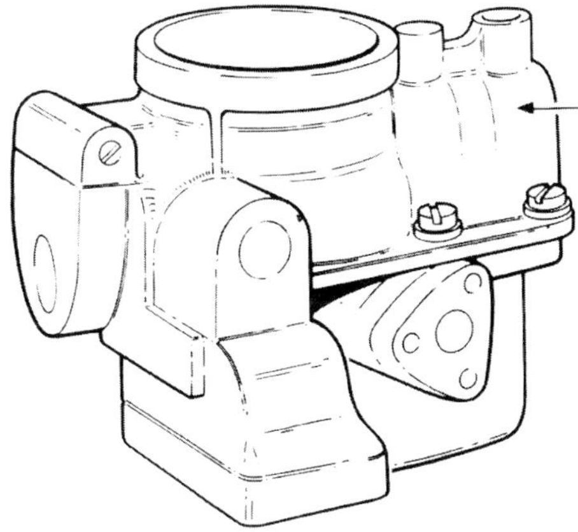

Fig. ***4.16*** Carburettor body

4.6 PLASTICS MATERIALS

Plastics materials are man-made or synthetic materials, many being produced from crude-oil products. When components are manufactured from plastics materials, at some stage heat and pressure are applied to the raw material.

The material becomes "plastic" and is moulded into the required shape. When the heat and pressure are removed the material stays in the moulded shape.

Most plastics materials can be classed in one of two categories

(1) **Thermoplastic materials**, which can be softened and reformed as often as they can be reheated ;

(2) **Thermosetting materials**, which cannot be softened and reformed after the initial forming operation; this group tends to be harder, more rigid and brittle compared with thermoplastic materials

Examples of some plastics materials are shown in Fig. 4.17 to 4.19.

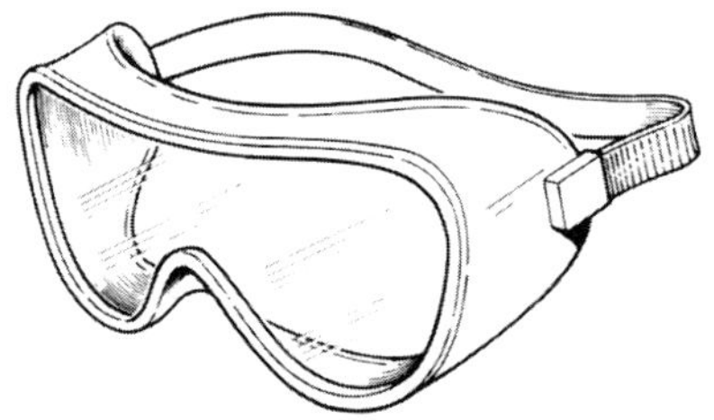

Safety goggles-cellulose acetate Fig. 4.17

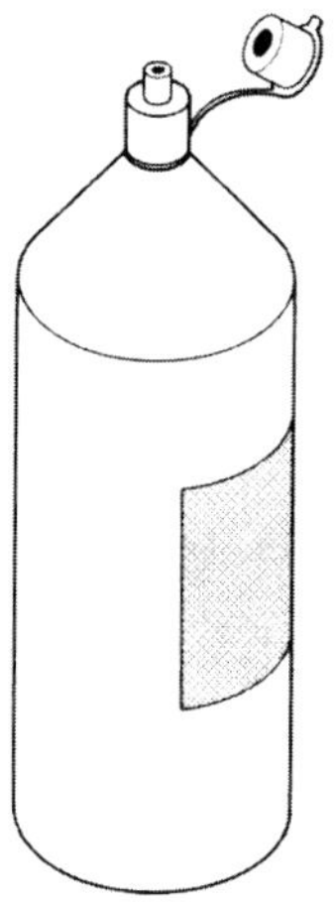

Polythene container Fig. 4.18

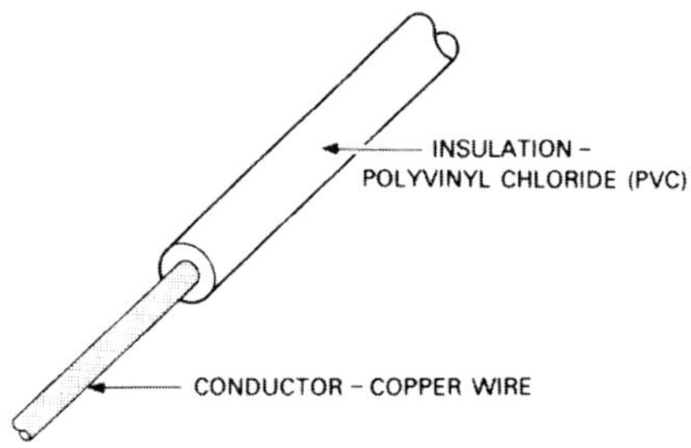

PVC on electric cable Fig. 4.19

4.7 SELECTING THE MATERIAL

Fig. 4.20, 4.21 and 4.22 show how similar materials (aluminium and its alloys) are used for the manufacture of different components, while Fig. 4.23 and 4.24 show how similar components (gears) may be manufactured from different materials. The question arises 'How is a material chosen for a component?'

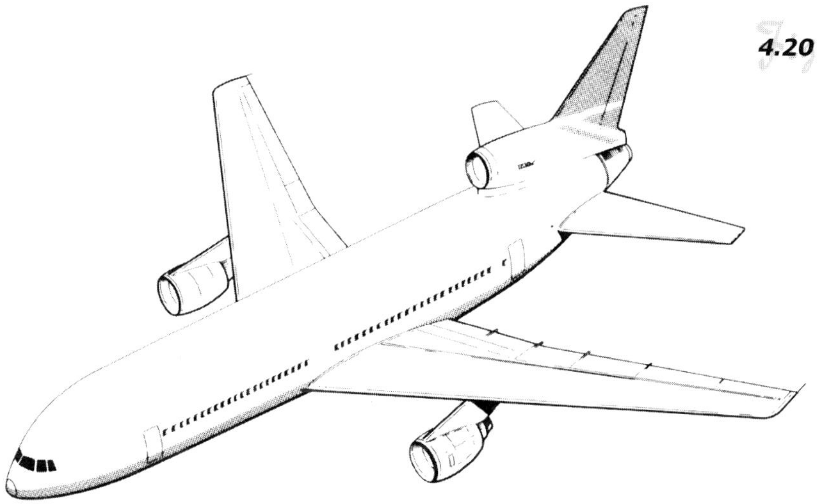

Fig. **4.20** Aeroplane

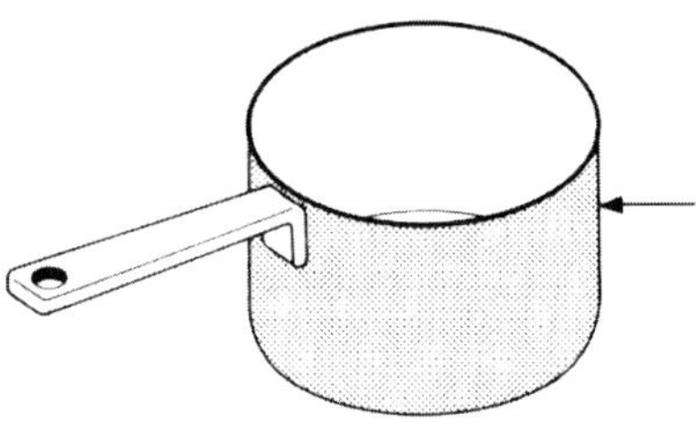

Fig. **4.21** Saucepan

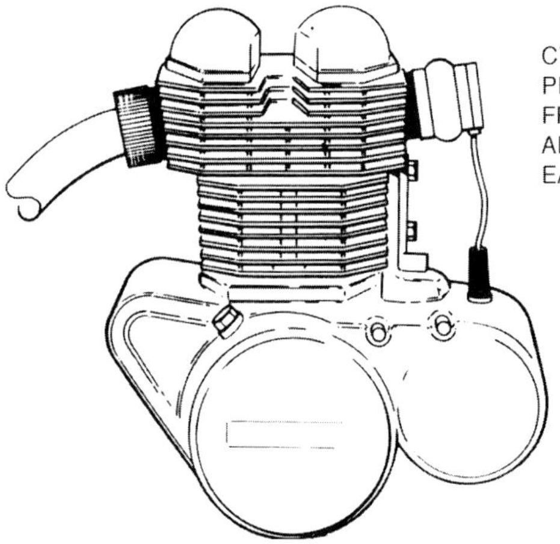

Motorcycle engine **Fig. 4.22**

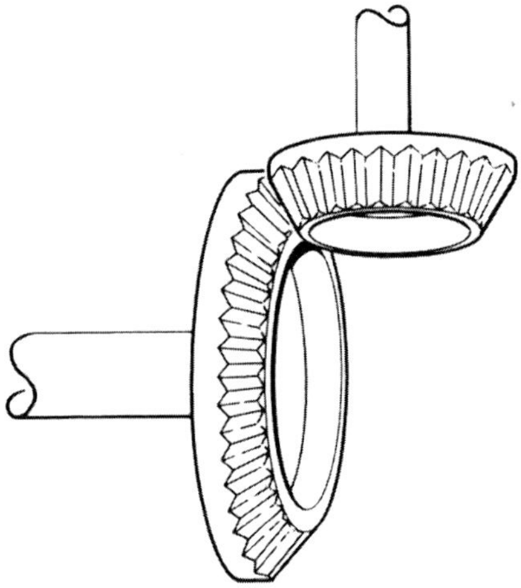

Nylon bevel gears-used in situation that require silent running with minimum lubrication **Fig. 4.23**

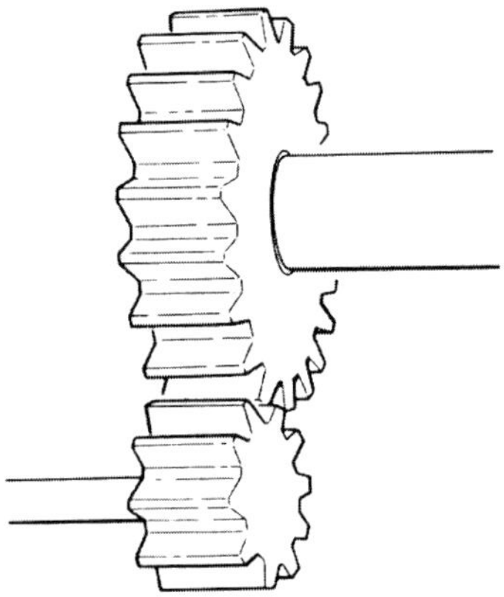

Cast-iron spur gear-used to drive lead screw mechanism in some lathes **Fig. 4.24**

VOCABULARY

mechanism ; 기구, 기계장치
component ; 구성요소, 구성품
factor ; 요인, 원인
plastic material ; 플라스틱 재료
insulator ; 절연체, 단열재
conductor ; 도체, 도선
behaviour ; 행실, 작용, 반응
conductivity ; 전도성, 도전율
mechanical property ; 기계적 특징
toughness ; 튼튼함, 질김
hardness ; 단단함, 경도
ductility ; 연성(延性)
malleability ; (금속의) 전성(展性)
withstand ; 견디다
stretch ; 잡아늘이다. 신축성
tensile strength ; 장력
shear strength ; 전단강도
subject ; 복종시키다, 맡기다.
impact strength ; 충격강도
ferrous metal ; 철금속
non-ferrous metal ; 비철금속
non-metallic ; 비금속성의
application ; 적용, 응용
high-carbon steel ; 고탄소강
mild steel ; 연 강, 저탄소강
axle ; 차축
corrosion ; 침식, 부식
brass ; 황동
bronze ; 청동, 청동제품
zinc-based material ; 아연기재료
man-made material ; 인공재료
crude-oil ; 원유
raw materials ; 원료
mould ; 주형, 금형
category ; 범주, 카타고리
thermoplastic ; 열가소성의
thermosetting ; 열경화성의
cellulose acetate ; 섬유소 인견
polythene ; 폴리에칠렌 (무색투명 한 합성수지의 =polyethylene)
visor ; 마스크, (자동차의)선바이저

명언 / 격언

Genius is one per cent inspiration and ninety-nine per cent perspiration.

— Edison

5

Manufacturing Technology Chapter

(제조기술)

5.1 INTRODUCTION

The processes used primarily to change the shape of metals are as follows:

(1) Casting

(2) Rolling

(3) Pressing

(4) Forging

(5) Plastics Molding

(6) Maching

5.2 CASTING

The body of the tap in Fig. 5.1 is made by a casting process. Note the shape of the internal passage ways, which allow the flow of water from the mains. To produce such internal features, casting is the easiest and cheapest method. It would be almost impossible to produce this component by any other method.

Casting, which is one of the oldest forms of shaping materials,

Chapter 5

involves injecting or pouring a molten metal into a mold, which is a similar shape to the required component. When the molten metal solidifies, a component similar in shape and size to the mould is obtained.

Water tap **Fig. 5.1**

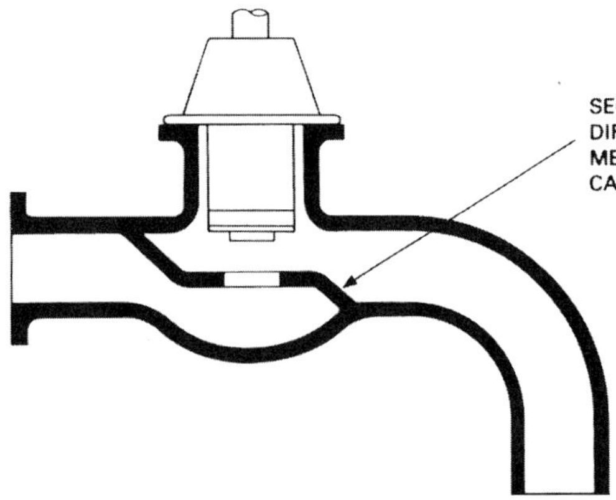

5.2.1 SAND CASTING

A sand casting is produced by pouring molten metal into a sand mould, which has been prepared with the aid of a suitable pattern.

When the metal has solidified, the sand mould is broken and the component removed. Hence a new mould is required for each casting. Fig. 5.2 shows this process.

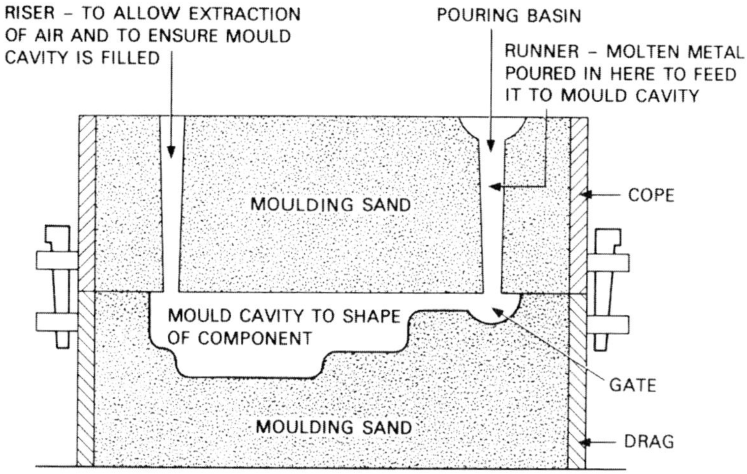

Fig. **5.2** Simple sand-casting

Cast iron is a material that is often used in the sand-casting process. The property that makes it so suitable is its ability, when molten, to fill and take up the intricate shapes of a sand mould. It is very suitable for large intricate components such as car cylinder blocks, machine tool beds and columns.

5.2.2 DIECASTING

The making of a sand mould for the casting of each component is a lengthy and relatively expensive process. The moulds used for diecasting are made from alloy steels, which means that they are permanent and not broken after each casting. Many thousands of components can be made from one set of moulds, or dies as they are usually called. A die set will consist of at least two parts so that they can be separated and the casting removed. This process is more suitable for metals that have a melting point that is rater lower than that of cast iron, such as zinc or aluminium-based alloys.

The metal can be poured into the dies in a similar way to sand casting, but for mass production of small and simple components, pressure diecasting processes are used. Here the molten metal is forced into the dies under pressure; this pressure tends to produce castings that have a better surface finish and greater dimensional accuracy. Fig. 5.3 and 5.4 show the principles of two pressure diecasting processes.

5.3 Hot chamber diecasting process- injection plunger moves vertically downwards and forces molten metal into component shape; Used for zinc-based diecasting alloys components such as carburetteors, fuel pumps and toy cars.

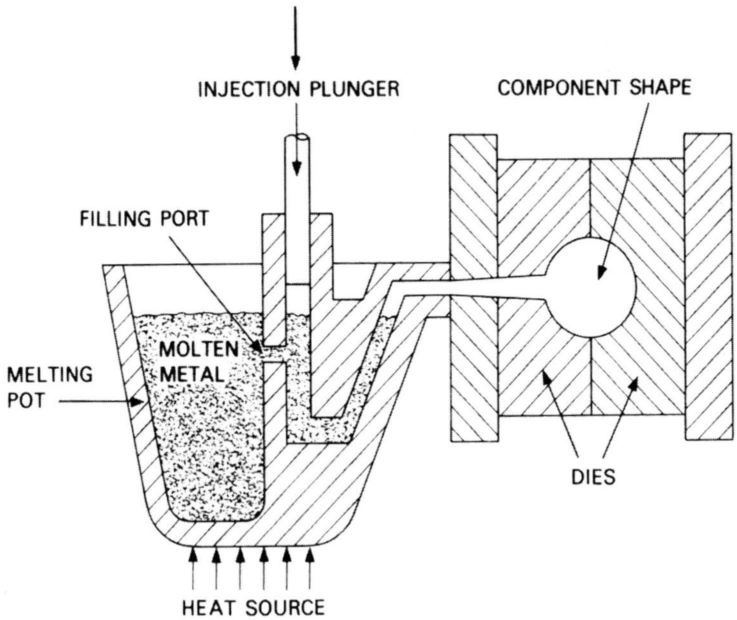

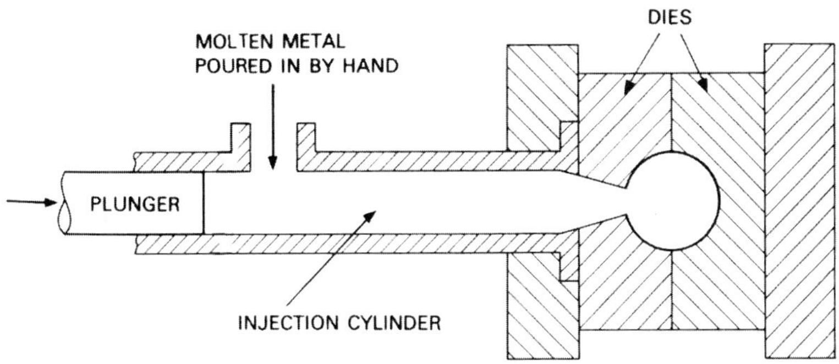

Fig. **5.4** Cold chamber diecasting process- used mainly for aluminium-based diecasting alloys for components such as crank cases, aircraft parts and engine pistons.

5.3 ROLLING

Look into the materials stores of any workshop and you will see that most of the metals in stock are either in bar or sheet form. Some typical sections of bar are shown in Fig. 5.5.

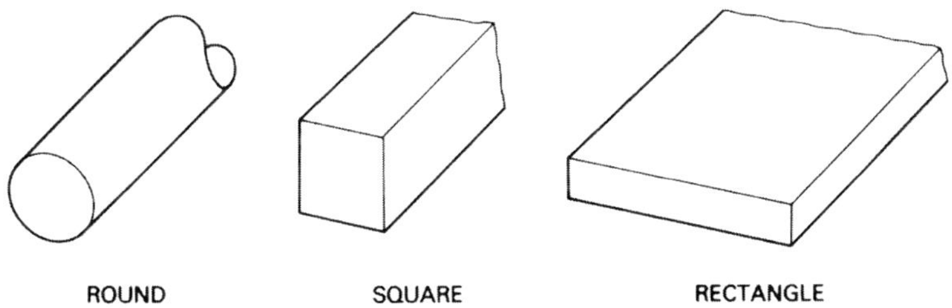

Fig. **5.5** Typical bar sections

When a metal such as steel is manufactured, it is initially cast into large ingots, which are roughly rectangular in shape. These ingots are then reduced to the sizes and shapes required by the engineering industry.

This reduction is achieved by **rolling.** The ingots are passed through a series of rolls (Fig. 5.6), which reduces the thickness, at the same time increasing the length. To make the steel easier to

Hot rolling two-high mill. **Fig. 5.6**

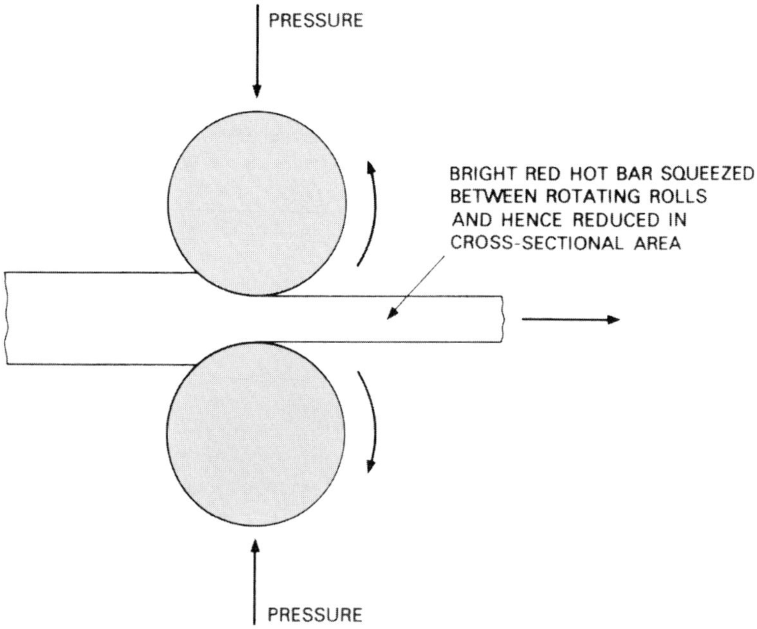

roll, it is heated to a bright red heat. The process of rolling metal that has been heated to a high temperature is called **hot rolling.**

Hot rolling has been used to produce the sections shown in Fig. 5.7. This process produces a black scaly surface on the component, which helps to stop the steel rusting in outdoor service. However, this type of surface is unsuitable for steel from which, say, car bodies are made. Such a component needs a smooth clean surface, and this can be achieved by **cold rolling.**

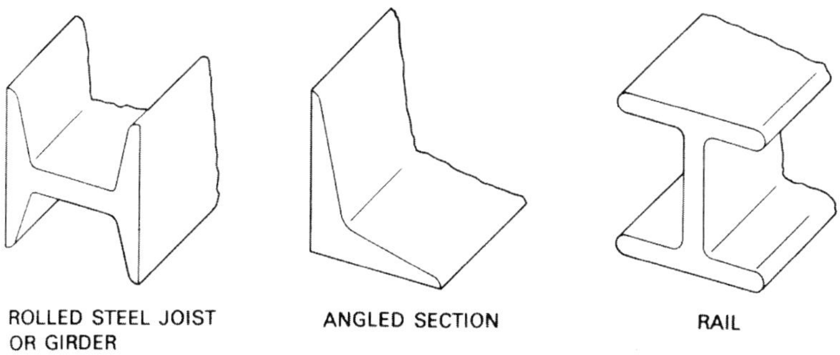

5.7 Typical steel shapes

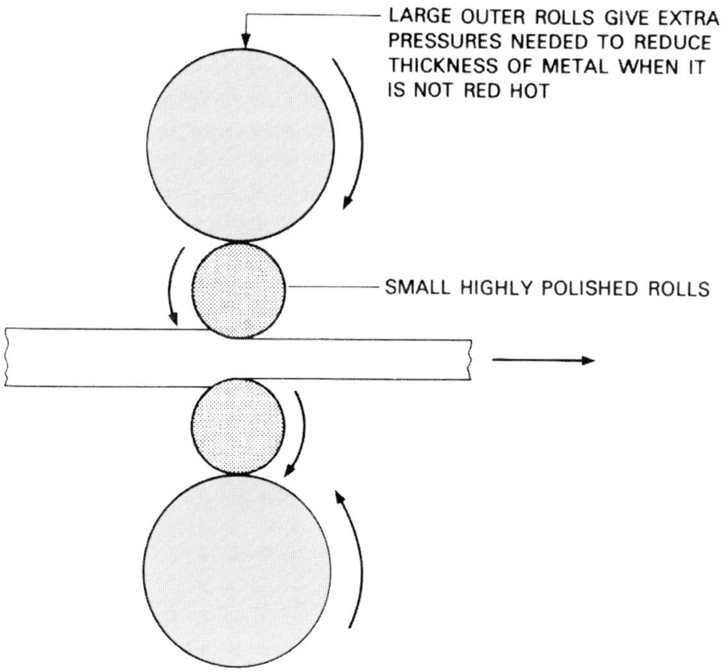

5.8 Four-high cold-rolling mill

The metal is first hot rolled near to the finished size, then pickled in acid to remove the scale. The final operation is to pass the metal through a type of rolling mill, which will provide the additional pressures to the cold metal. The rolls are highly polished and this finish is reproduced on the surface of the metal. The process is shown in Fig. 5.8 and a typical component in Fig.

Car body panels Fig. 5.9

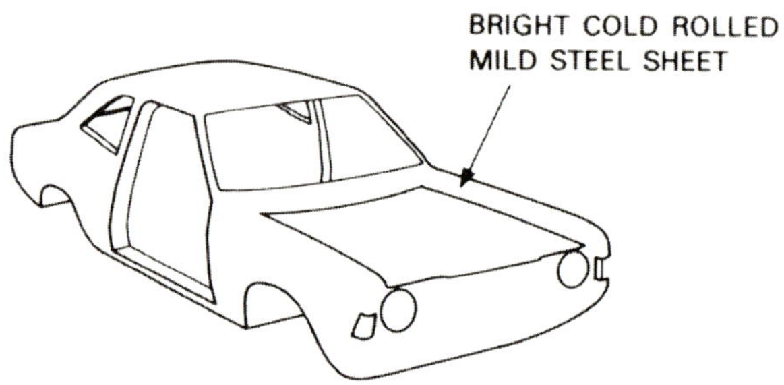

5.4 PRESSING (SHEET-METAL FORMING)

The articles shown in Fig. 5.10 are completely different in size, shape and sometimes in material. The thing they have in common is the process by which they have been manufactured. This is **pressing**, from either sheet metal or strip. A hydraulic press used for the operation are shown in Fig. 5.11.

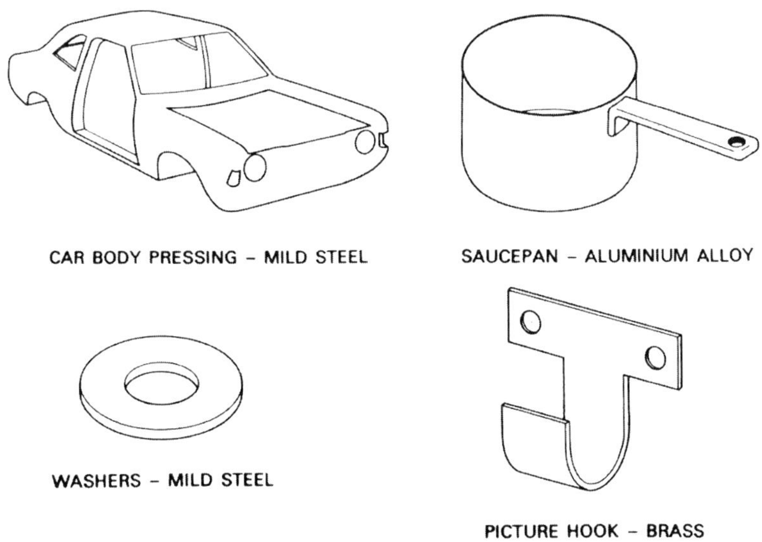

5.10 Hand press

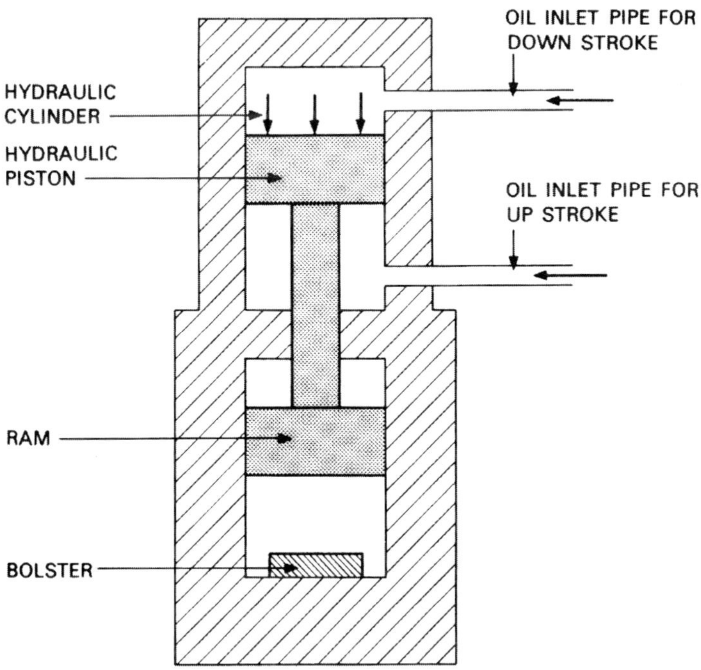

5.11 Principle of hydraulic press

5.4.1 BLANKING AND PIERCING

One type of presswork is punch holes and cut out shapes from sheet metal. A simple component production and the principles involved are shown in Fig. 5.12.

Fig. 5.12 Principle of follow-on blank-and piercing tools

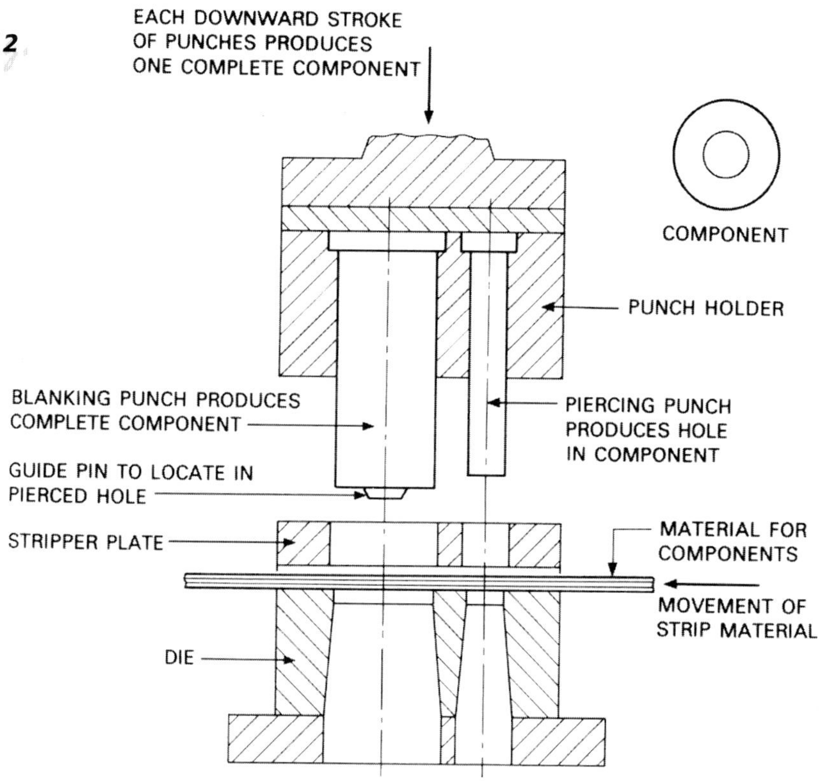

5.4.2 BENDING

Simple components such as picture hooks and pipe clips (Fig. 5.13) can be produced by bending. The blanks for the bending operation may have been previously cut or blanked out using the methods outlined in section 5.4.1 The bending tools themselves are usually of simple construction, as shown in Fig. 5.14.

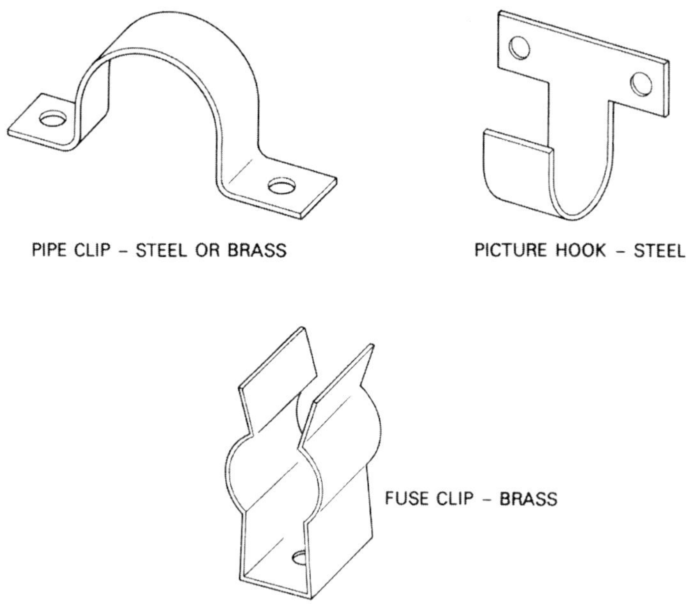

Fig. 5.13 Components produced by bending

Fig. 5.14 Principle of bending -and simple vee-type bending tools

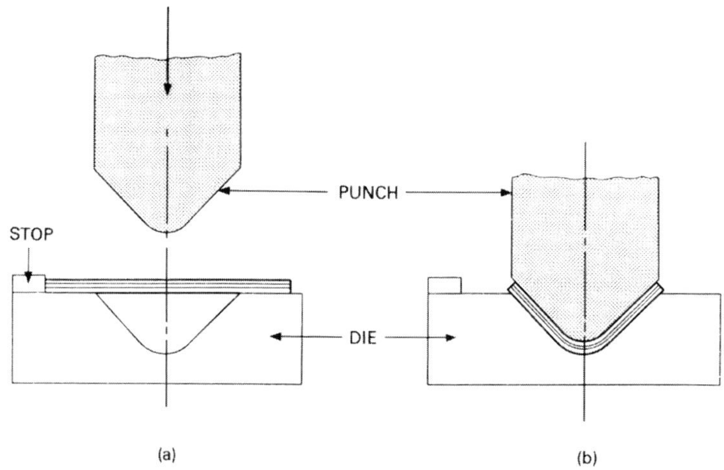

5.4.3 DRAWING

Car bodies, saucepans, brake drums, trays and cylindrical containers are all examples of components produced from flat sheet metal that has been pressed into a dish or cup form. Such a pressing operation is called drawing.

The stages of the process are shown in Fig. 5.15. Notice that the punch draws the material over the edge of the die to form the cup.

Fig. 5.15 Principle of deep drawing

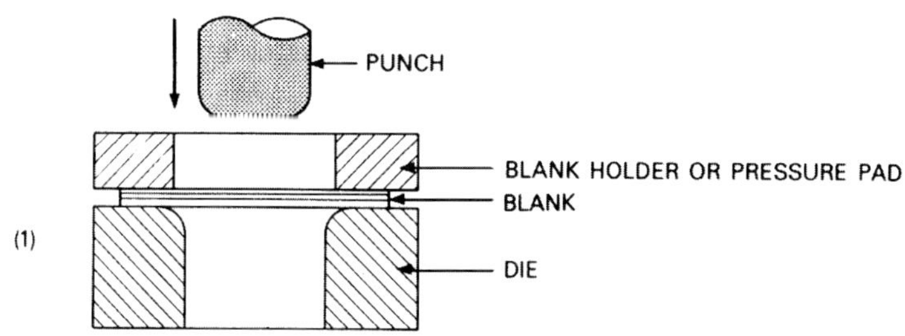

(2)

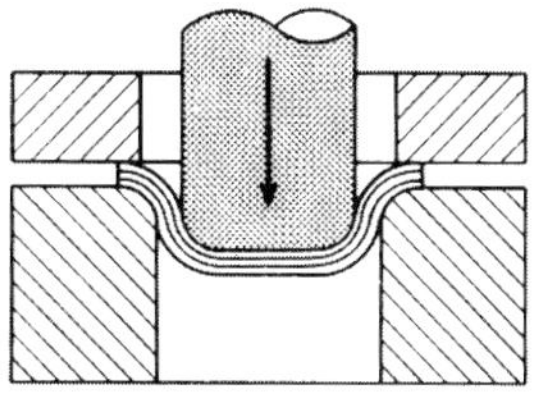

(3)

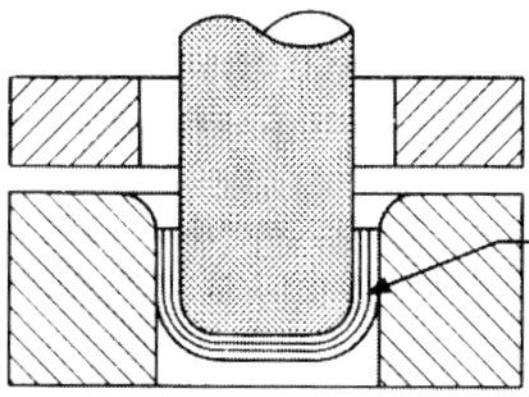

5.5 FORGING

This process shapes metals by hammering or squeezing them to the shapes required. The metal is often heated to make the operation easier. Hand forging (Fig. 5.16), as performed by a blacksmith, is only used for small components produced in small numbers. For batch quantities and large sized components power hammers or hydraulic presses are used.

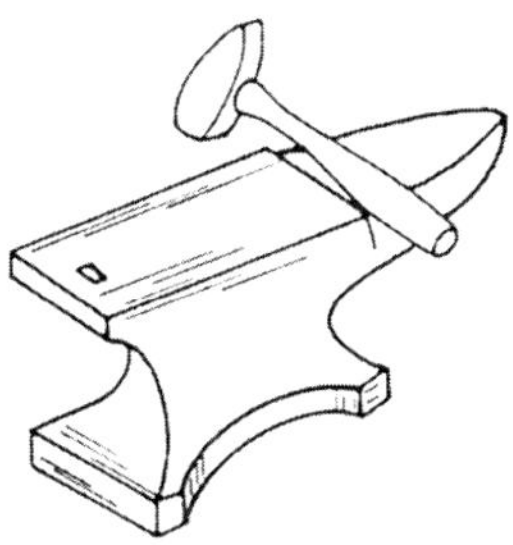

5.16 Hand forging.

5.6 PLASTICS MOULDING

There are two main groups of plastics materials. They are as follows

(1) Thermosetting plastics

(2) Thermoplastics

5.6.1 THERMOSETTING PLASTICS

Components made from these materials are produced by compressing or squeezing the plastics material, in a powder form, between heated dies or patterns, which have the shape of the component cut into them (Fig. 5.17). Under the action of heat and pressure, the powder becomes plasticised, takes up the die form and becomes a solid component.

Compression moulding *Fig.* **5.17**

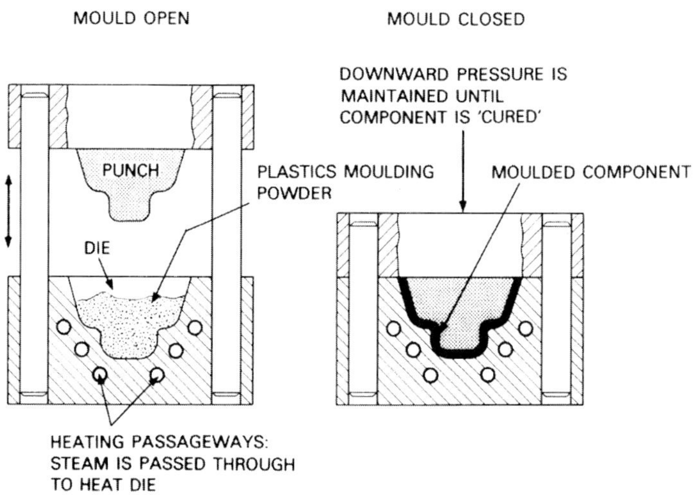

5.6.2 THERMOPLASTICS

Components produced from thermoplastic materials can be manufactured in a similar way to those made from thermosetting materials.

However, for large-quantity production it is more usual to use an injection-moulding machine of the type shown in Fig. 5.18. Another process used to shape plastics sheet is vacuum forming.

The principle is shown in Fig. 5.19. A plastics sheet is placed over the mould, clamped to form an air-tight seal, and heated to just below its melting point. Air between the sheet and the mould is evacuated by a pump. The outside atmospheric pressure is sufficient to force the sheet into the mould form, where on cooling it takes up the mould form.

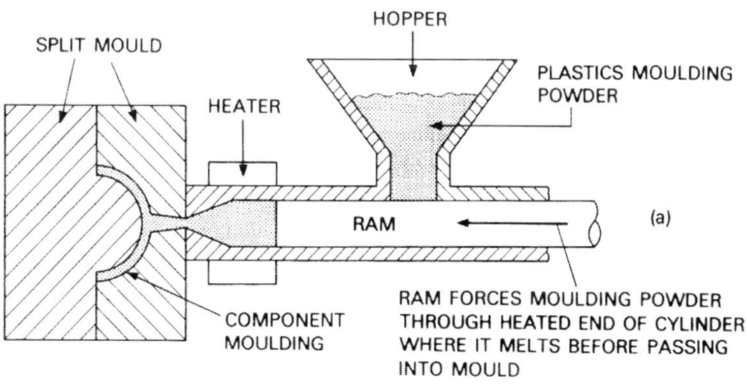

5.18 Injection moulding

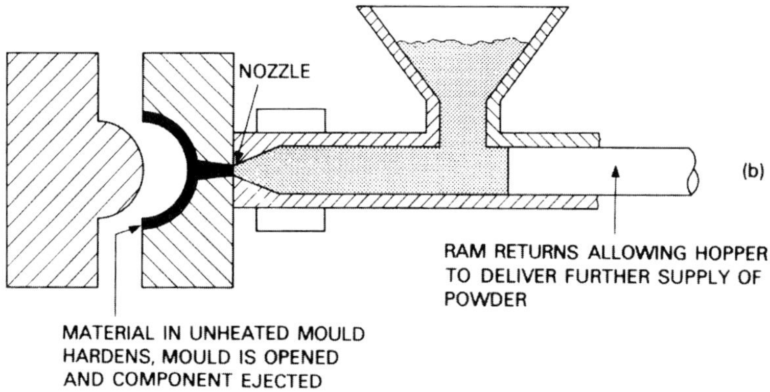

Vacuum forming *Fig.* **5.19**

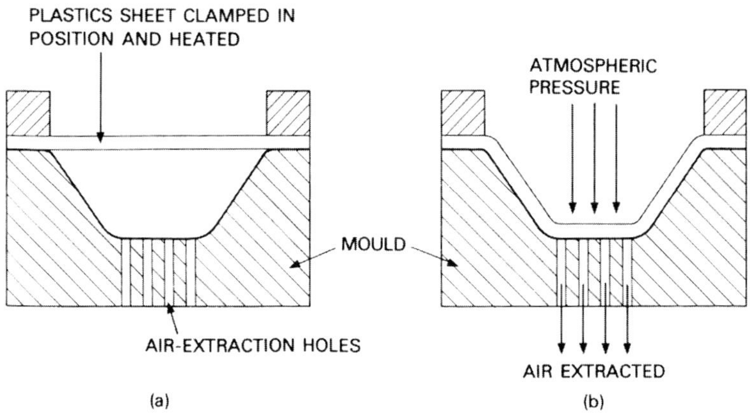

5.7 MACHINING

Study the components shown in Fig. 5.20 and 5.21. You will notice that these components could be one or more of the processes we have previously considered. Each of these components could also have been made by **machining** them a solid piece of material.

Making components by machining involves cutting away material to leave the required shape of the component.

5.7.1 FACING AND TURNING

Facing and turning are operations that are performed on the lathe. Usually, in the operation of a lathe, the piece to be worked is held and turned by a rotating vise called a chuck, and the metal is removed by a nonrotation cutter. When the cutter moves perpendicular to the axis of the workpiece, the operation is known as facing; when it moves parallel to the workpiece axis, the operation is called turning (Fig. 5.20).

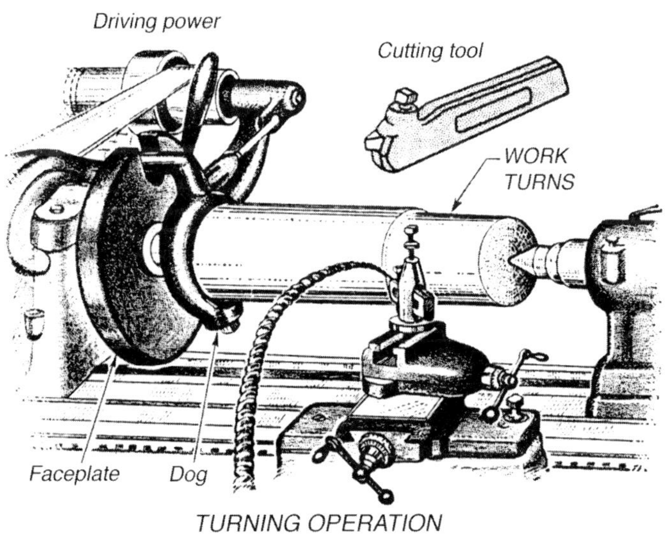

5.20 Facing and turning.

5.7.2 MILLING

Milling machine operations are among the comprehensive accomplished in a machine shop. Gear cutting and shaping are perhaps the most common jobs done on a milling machine. The milling machine cutter is usually shaped to the desired appearance of the cut required and it turns until the cut is completed. The work is then moved to the next position for the succeeding cutting operation (Fig. 5.21).

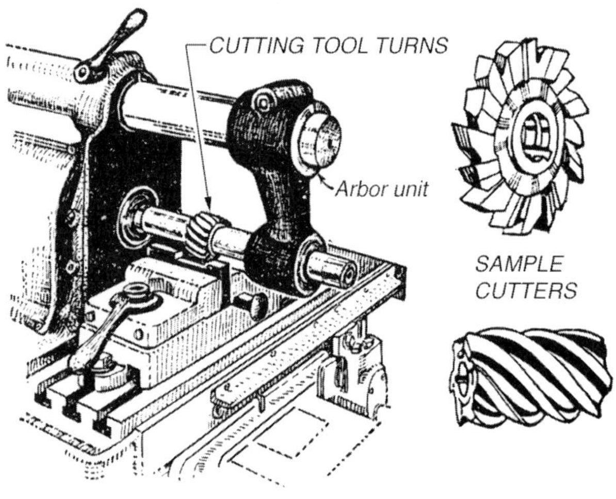

Horizontal milling machine *Fig.* **5.21**

5.7.3 DRILLING

Drilling may be accomplished on an lathe or a milling machine, but the drill press is the machine designed for this operation. In this machine the tool turns and can be raised and lowered. while the work is held in a fixed position (Fig. 5.22).

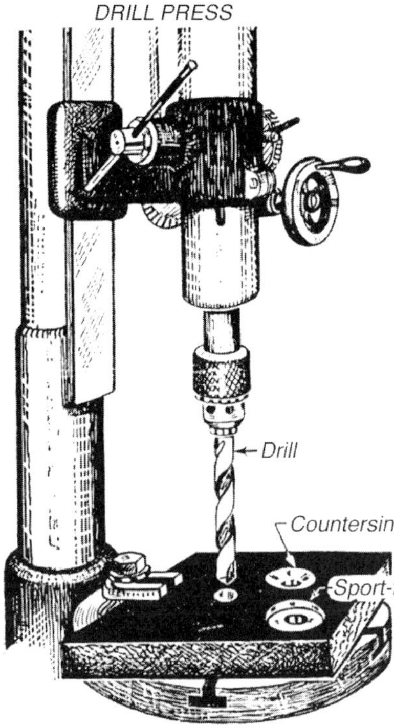

5.22 Drill press.

5.7.4 REAMING

Drilled holes are rarely absolutely straight, cylindrical, or accurate in size. To "true" them they are reamed out by a tool that removes the riflings left by drill. This operation can be performed on any machine that can be used for drilling(Fig. 5.23)

Reaming. Fig. 5.23

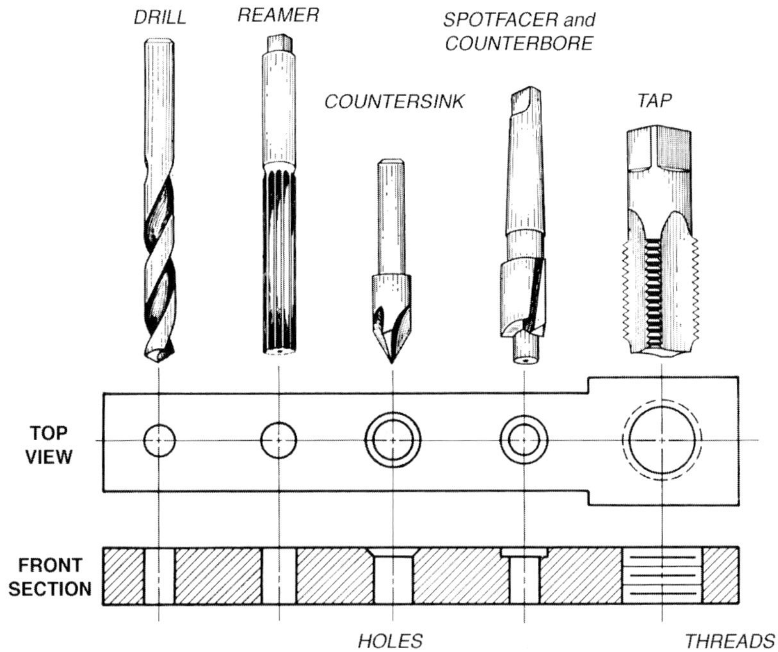

5.7.5 GRINDING

This operation, intended to bring surfaces to a fine finish, is accomplished by the removal of a very small amount of the material as both the work and the grinder wheel turn at high speed, or as the grinder turns while the work is moved only at intervals. Grinding machines fall into four groups known as surface, cylindrical, tool, and centerless grinders(Fig. 5.24).

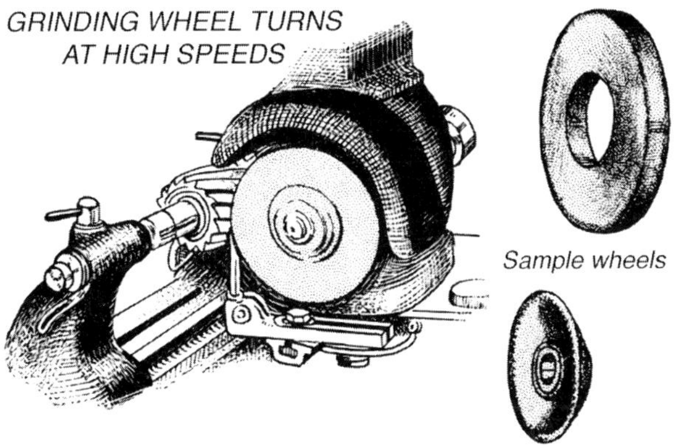

5.24 Grinding.

Chapter 5

VOCABULARY

technology ; 생산[공업]기술,
achieve ; 목적을 이루다
require ; 요구하다
process ; 과정, 순서
suitability ; 적당, 적합
tap ; 주둥이, 수도꼭지
solidify ; 응고 시키다, 굳히다.
molten metal ; 용융금속
intricate ; 뒤엉킨, 복잡한
diecasting ; 압력주조, 다이주물
rectangle ; 직사각형
initially ; 처음의, 최초의,
ingot ; 주괴. 금괴로 만들다.
hot rolling ; 열간압연
scaly ; 비늘이 있는, 더러운
cold rolling ; 냉간압연
pickle ; 희박산
수로 씻다.
rolling mill ; 압연기
polish ; 닦다, 광택
article ; 기사, 논설
blanking ; 블랭킹, 천공
piercing ; 피어싱, 꿰뚫는, 천공
picture hook ; 그림 걸쇠
pipe clip ; 파이프 조임틀
previously ; 이전에, 미리
saucepan ; 소스냄비
tray ; 쟁반모양의 접시
facing ; (선반에서) 단면절삭
turning ; 선삭
lathe ; 선반
chuck ; 척, (선반의 물림쇠)
perpendicular ; 수직의
reaming ; (구멍) 넓히기
rifling ; 강선을 붙임, 선조
milling ; 평삭
comprehensive ; 광범위한, 넓은
appearance ; 출현, 외관
grinding ; 빻기, 찧기
interval ; (시간, 장소의)간격

명언 / 격언

Early to bed and early to rise,
make a man healthy, wealthy and wise.

— *Benjamin Franklin*

6

Welding Chapter

(용접)

6.1 INTRODUCTION

Welding is an important industrial process used for joining all kinds of metal parts. It is a process in which the workpieces are melted along a common edge or surface so that their molten metal, along with a molten filler metal, forms a common bond when the materials solidify.

Many parts that were formerly cast or forged are now being manufactured or constructed more easily, less costly, and with less weight by using metal fabrication techniques.

More than 40 types of welding processes are used in industry today. However, only three are commonly used and represent the vast majority of all welding done. These include **arc welding**, **gas welding**, and **resistance welding.**

Chapter 6

6.2 OXY-ACETYLENE GAS WELDING

A set of **oxy-acetylene gas welding** equipment is shown in Fig. 6.1 and its principle of operation in Fig. 6.2. The cylinder controls are used to set the correct **gas pressure**; the torch controls are used to adjust the flame to a setting suitable for the particular application. Only qualified craftsmen should attempt to set gas welding equipment, and care is required in the shutting-down as well as the setting-up.

Oxy-acetylene welding set *Fig.* **6.1**

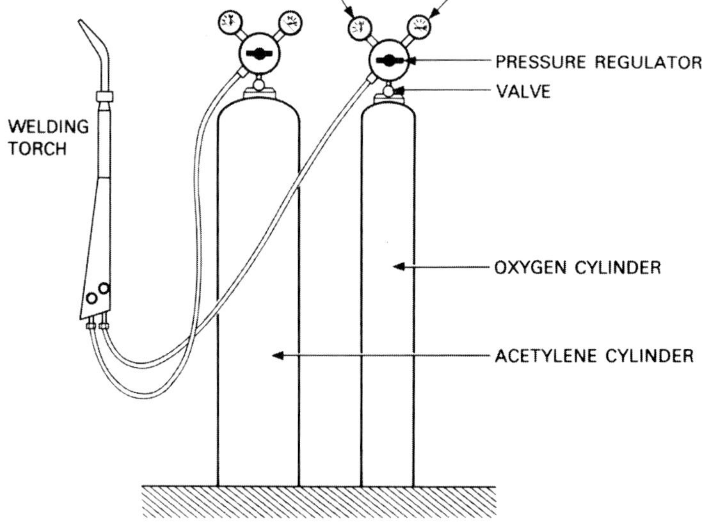

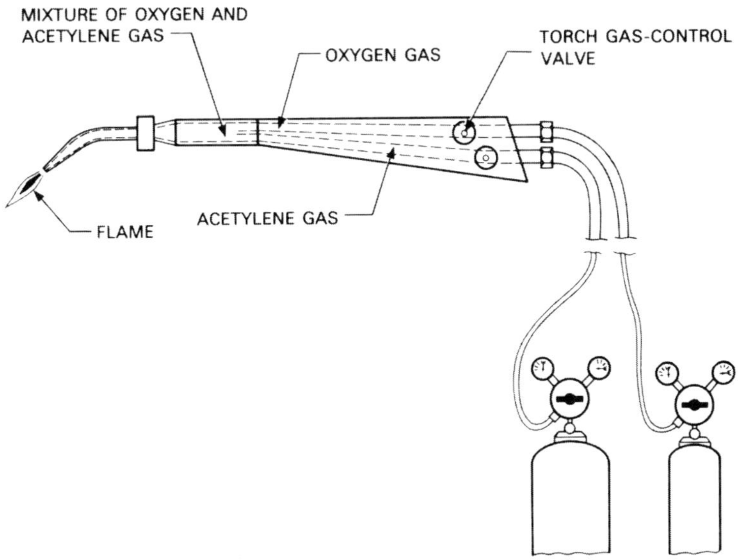

Fig. **6.2** Gas welding -principle of operation

The temperature of the oxy-acetylene flame is about 3200℃, which is sufficient to melt the edges of steel components and mild steel filler rods. The production of a gas-welded joint is shown in Fig. 6.3, with some different techniques in Fig. 6.4 and 6.5.

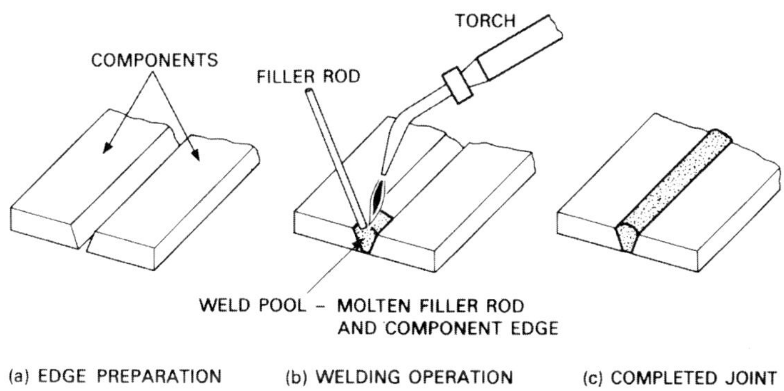

Fig. **6.3** Gas welding

Leftward welding. Fig. 6.4

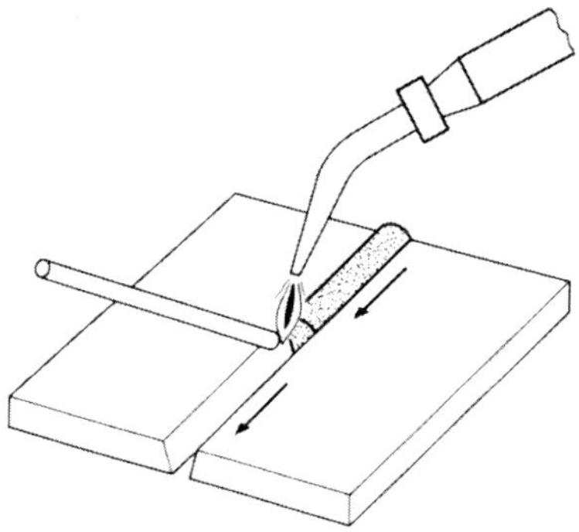

Rightward welding. Fig. 6.5

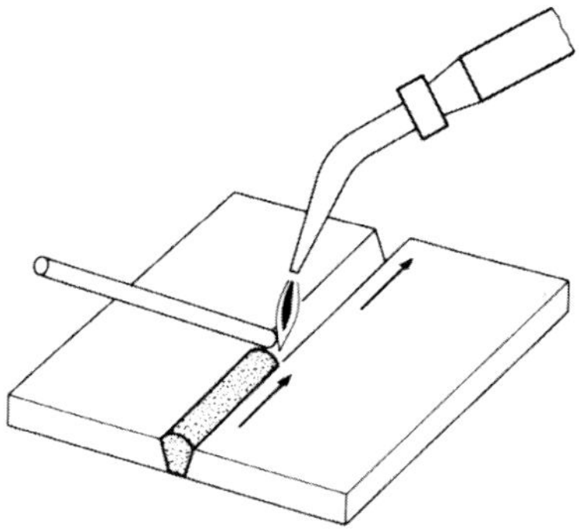

Different forms of edge preparation are shown in Fig. 6.6. Whenever someone is using gas-welding equipment, or even watching, the following protective clothing is necessary.

(1) tinted goggles to protect eyes from the glare and sparks.

(2) suitable aprons, gloves and footwear for protection from the heat and sparks.

METAL THICKNESS		FILLER ROD DIAMETER		EDGE PREPARATION
mm	in.	mm	in.	
3 - 5	$\frac{1}{8}$ - $\frac{3}{16}$	3 - 5	$\frac{1}{8}$ - $\frac{5}{32}$	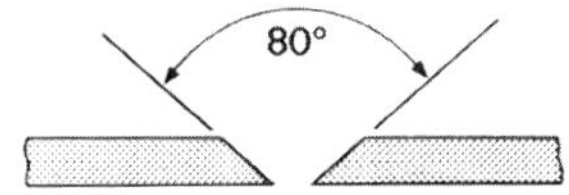
5 - 8	$\frac{3}{16}$ - $\frac{5}{16}$	3 - 5	$\frac{1}{8}$ - $\frac{5}{32}$	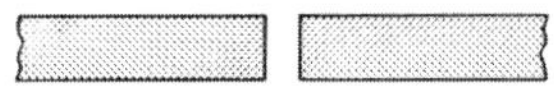
8 - 16	$\frac{5}{16}$ - $\frac{5}{8}$	5 - 6	$\frac{5}{32}$ - $\frac{1}{4}$	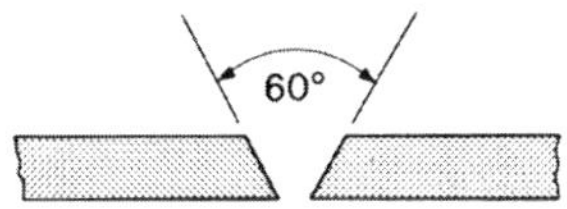
16^{+}	$\frac{5}{8}^{+}$	6	$\frac{1}{4}$	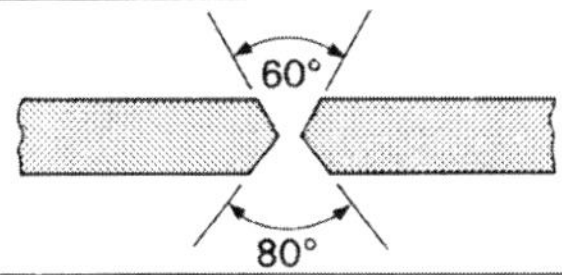

Fig. 6.6 Edge preparation

6.3 ARC WELDING

Arc welding is a welding process more suitable for joining thicker sections. A typical set-up is shown in Fig. 6.7. An electric circuit is broken by the gap between the electrode and the work.

Arc-welding set-up *Fig.* **6.7**

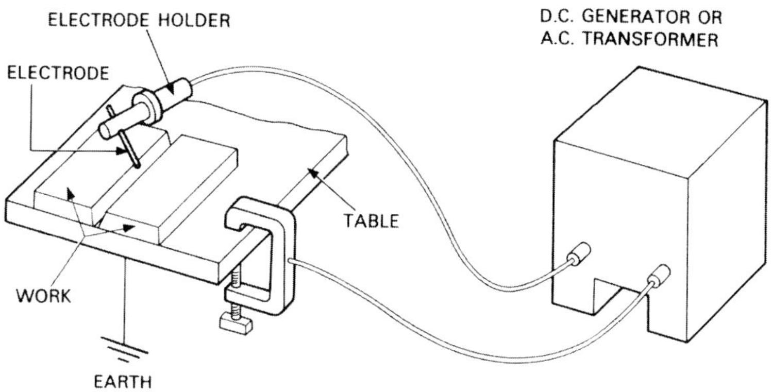

In operation an arc bridges this gap and generates heat with a temperature of about 6000℃. The heat melts the component edges and the steel filler rod, which is located in the centre of the electrode.

The outside cover of the electrode acts as a gaseous flux (Fig. 6.8). On completion of the welding, a slag covering has be chipped away.

Arc-welding action *Fig.* **6.8**

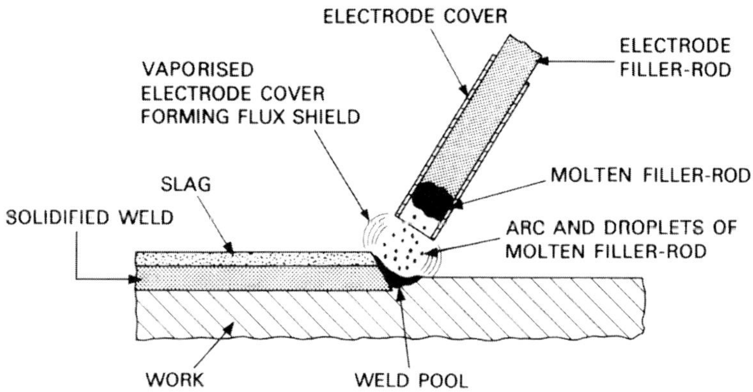

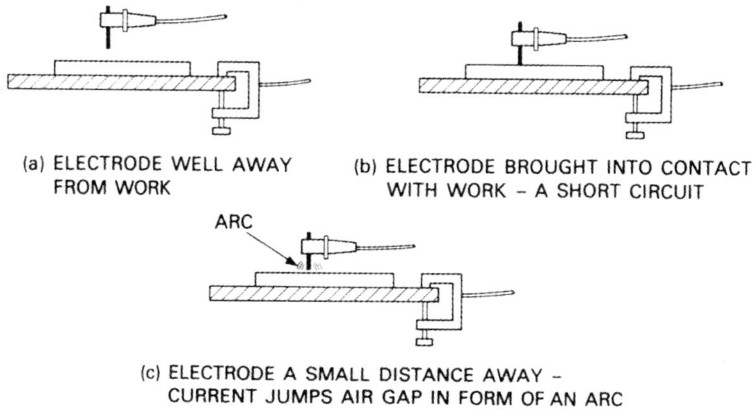

Fig. 6.9 Striking an arc

The procedure for striking an arc is shown in Fig. 6.9 With arc welding, suitable clothing is even more important than with gas welding.

The arc produces ultraviolet rays and a special tinted face visor must be used. Gas welding goggles will not provide suitable protection. Complete face coverage is necessary for protection from the heat and sparks. Such is the danger from ultraviolet ray that screens must be placed around all areas where arc welding is taking place, so protecting anyone in the vicinity.

Suitable aprons and gauntlets are also necessary. Adequate ventilation is required to prevent the fumes that result from the operation becoming a hazard. Fig. 6.10 shows examples of protective clothing.

Protective clothing for arc welding *Fig.* **6.10**

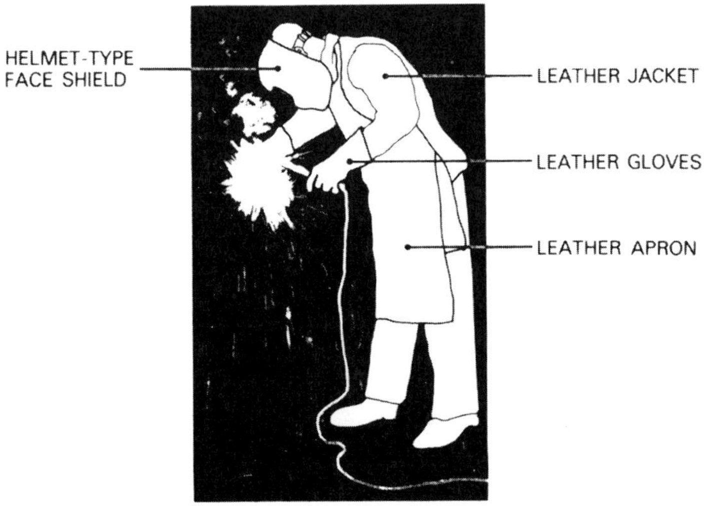

6.4 RESISTANCE WELDING

Resistance welding uses the principle that a heavy electric current flowing through a resistance become hot. Spot welding is the most common application of resistance welding. Component that are to be joined together are held between two copper electrodes and a heavy current passed through them for a set period of time (Fig. 6.11)

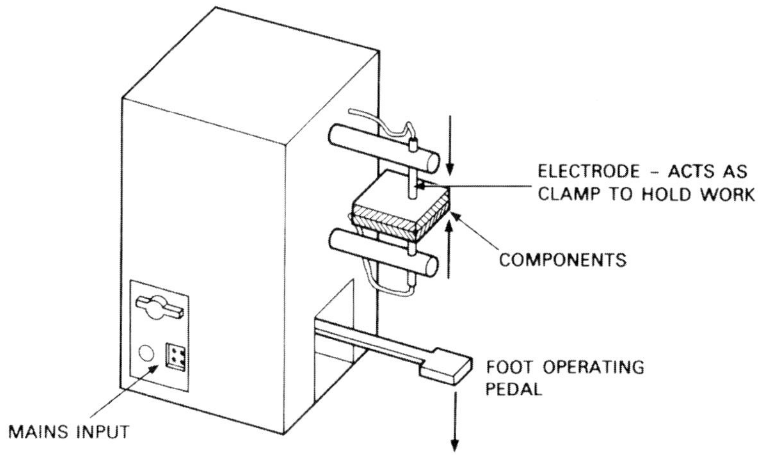

6.11 Resistance(spot) welding unit

The metal components form a resistance to current flow, causing a rise in temperature, which, combined with the clamping pressure, is sufficient to form a nugget-shaped weld (Fig. 6.12a). Long components are often joined with a series of spot welds (Fig. 6.12b)

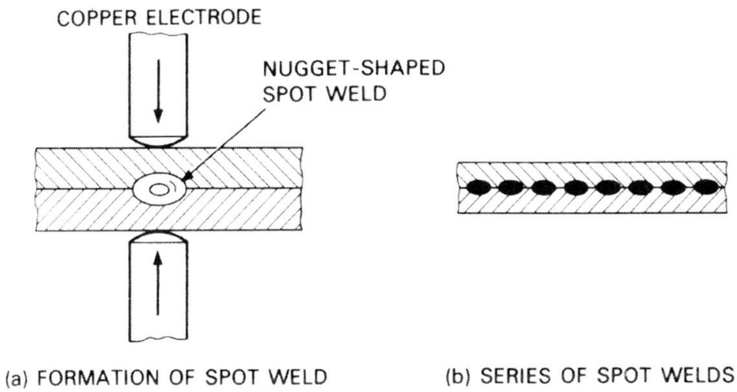

6.12 Spot welding

Chapter 6

VOCABULARY

industrial ; 상업상의, 공업상의
workpiece ; 제조공정에 있는 소재
solidify ; 응고시키다. 굳히다.
fabrication ; 제작, 구성, 제조
vast ; 거대한, 막대한
majority ; 대부분, 과반수
arc welding ; 아크 용접
gas welding ; 가스 용접
resistance welding ; 전기저항용접
qualified craftsmen ; 유자격 기 능공
hazard ; 위험
spot welding ; 점용접
nugget ; 금괴, 천연금속 덩어리
flux ; 흐름, 유동
slag ; 용재
chip ; 도막, 조각
ultraviolet ; 자외선의, 자외선
coverage ; 적용범위, 보상
vicinity ; 접근, 근처 부근
gauntlet ; 가죽으로 만든 긴 장갑
adequate ; 충분한
ventilation ; 공기의 유통, 환기
fume ; (유해, 불쾌한)김, 연기
setting-up ; 조립
tinted goggles ; 엷은색의 안경
glare ; 섬광, 번쩍이는 빛

명언 / 격언

From birth to age eighteen, a girl needs good parents, from eighteen to thirty-five she needs good looks, from thirty-five to fifty-five she needs a good personnality, and from fifty on she needs cash.

— *Sophie Tucker*

Chapter 7

Safety in the Shop

(공장 안전)

7.1 ACCIDENTS

Accidents do not happen, they are caused. There is not a single accident that could not have been prevented by care and forethought on somebody' s part. Accidents can and must be prevented.

They cost millions of pounds every year in damage and loss of premises, plant and lost business. They cost millions of lost working hours each year, but these are of little importance compared with the immeasurable cost in human suffering.

In every eight-hour shift nearly 100 workers are the victims of industrial accidents. Many of them will be blinded, maimed for life, or confined to a hospital bed for months. At least two of them will die. Fig. 7.1 shows the main causes of accidents.

Chapter 7

Fig. 7.1 Causes of industrial accidents

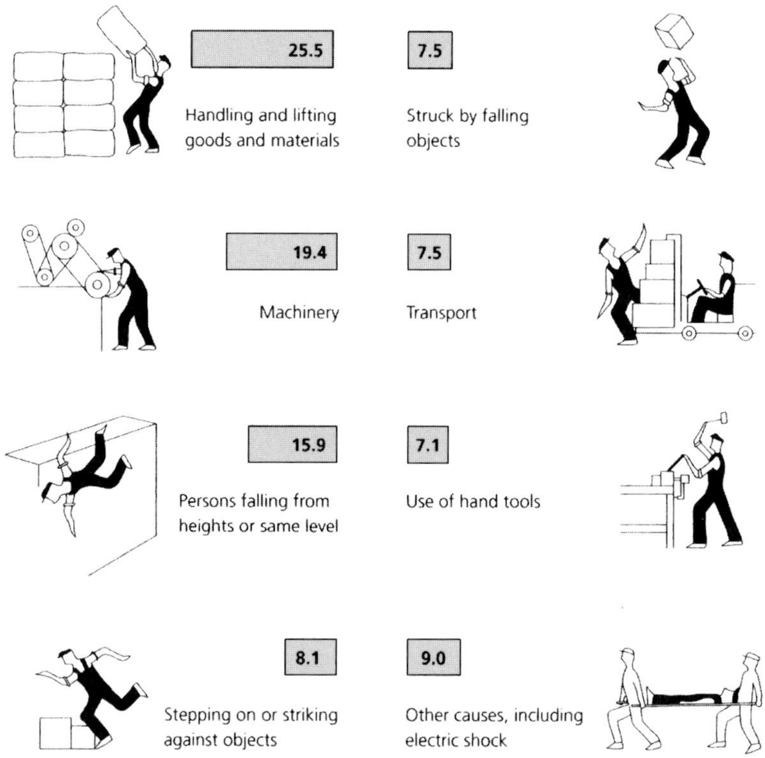

7.2 PROTECTIVE CLOTHING

For general workshop purposes the **boiler suit** is the most practical and the safest form of body protection. However, to be completely effective, certain precautions must be taken (Fig. 7.2).

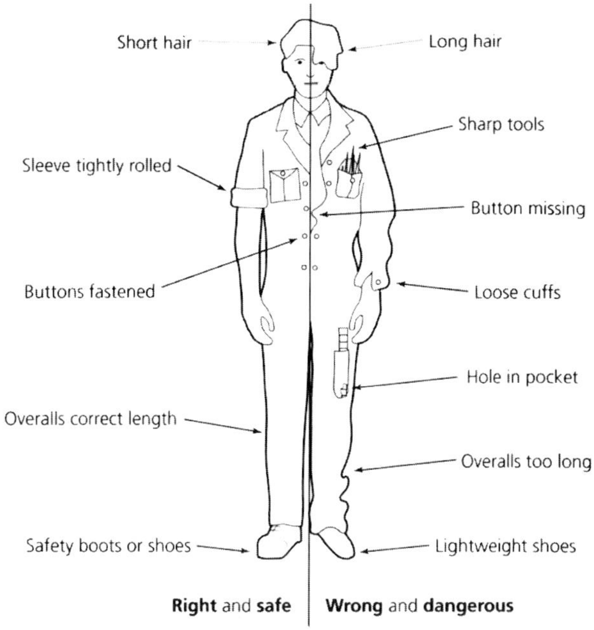

7.2 Safe dress

7.2.1 SHARP TOOLS

Sharp tools protruding from the breast pocket can cause severe wounds to the wrist. Since the motor nerves of the fingers are near the surface in the wrist, these wounds can paralyse the hand and fingers.

7.2.2 BUTTON MISSING

Since the overall cannot be fastened properly, it becomes as dangerous as any other loose clothing and is liable to be caught in moving machinery.

7.2.3 LOOSE CUFFS

Not only are loose cuffs liable to be caught up like any other loose clothing, they may also prevent the wearer from snatching his or her hand away from a dangerous situation.

7.2.4 HOLE IN POCKET

Tools Placed in the pocket can fall through onto the feet of the wearer. This may not seem potentially dangerous, as the feet should be protected by stout shoes, but it could cause an accident by distracting attention at a crucial moment.

7.2.5 OVERALLS TOO LONG

Excessively long overalls can cause falls, particularly when negotiating stairways.

7.2.6 LIGHTWEIGHT SHOES

The possible injuries associated with lightweight and unsuitable shoes are:

- Severe puncture wounds caused by treading on sharp objects.
- Crushed toes caused by falling objects.
- Damage to the Achilles' tendon due to insufficient protection around the heel and ankle.

In addition to body protection, it is necessary to protect the head, eyes, hands and feet. Let's now consider suitable protective clothing.

7.3 HEAD PROTECTION

Long hair is a very real hazard in any workshop.

- Long hair is liable to be caught in moving machinery, particularly drilling machines and lathes. When the hair and scalp are torn away, the resulting wound is both extremely painful and dangerous. Brain damage may also occur.
- Long hair is also a health hazard, as it is almost impossible to keep it clean and free from infection in the workshop environment.

If your hair becomes entangled in a machine (Fig. 7.3), you are very likely to be scalped a very painful and serious injury. If a fitter or machinist persists in retaining a long hairstyle in the interests of fashion, then the hair must be contained within a close-fitting cap. This also helps to keep the hair and scalp clean and healthy.

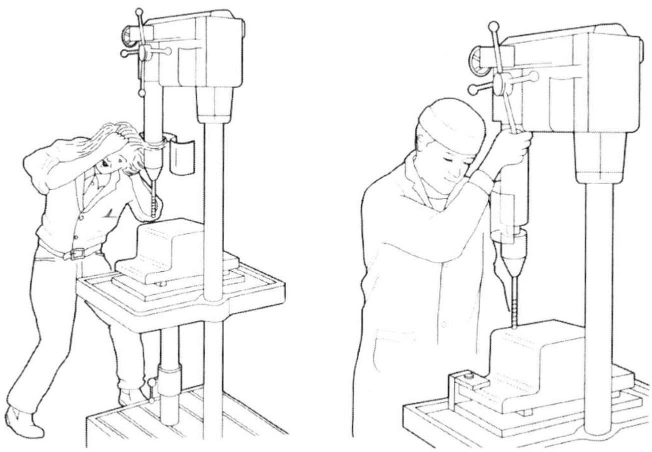

7.3 Hazards of long hair.

7.4 EYE PROTECTION

Although it is possible to walk on a wooden leg, nobody has ever seen out of a glass eye. Therefore eye protection is possibly the most important safety precaution to take in the workshop. Eye protection is provided by wearing suitable goggles or visors(Fig. 7.4).

When welding, special goggles have to be worn with coloured lenses to filter out harmful rays. Gas welding goggles are not suitable when arc welding. Eye injuries fall into three main categories:

- Pain and inflammation due to abrasive grit and dust getting between the lid and the eye.
- Damage caused by exposure to ultraviolet radiation (arc welding) and to high- intensity visible radiation. Particular care is required when using laser equipment.
- Loss of sight due to the eyeball being punctured or the optic nerve being severed by flying splinters of metal (swarf, or by the blast from a jet of compressed air).

Fig. 7.4 Safety goggles and visor: (a) transparent goggles suitable for machining operations. (b) complete plastic face visor for protection against chemical splashes.

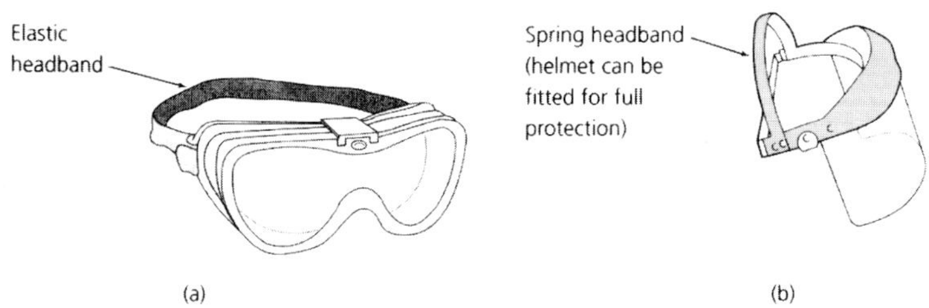

7.5 HAND PROTECTION

An engineer' s hands are in constant use and they run the risk of handling dirty, oily, greasy, rough, sharp, brittle, hot and maybe toxic and corrosive materials.

Gloves and palms in a variety of styles and materials are available to protect the hands, whatever the nature of the work.

Gloves are sometimes inappropriate, e.g. for working precision machines, but hands still need to be protected from oil and grime, though not cuts and abrasions, by rubbing them in a barrier cream before starting to work.

This is a mildly antiseptic, water-soluble cream which fills the pores of the skin and prevents the ingress of dirt and subsequent infection. The cream is easily removed by washing, which carries away the dirt and removes sources of infection.

7.6 FOOT PROTECTION

Unsuitable footwear should always be discouraged. It is not only false economy, but extremely dangerous to wear lightweight casual or sports shoes in the workplace. They offer no protection from crushing or penetration.

In safety footwear, protection is provided by a steel toecap (inside the boot or shoe) which conforms to a strength specification in accordance with BS 1870. Safety footwear is available in a wide range of styles and prices. It can be attractive in appearance and comfortable to wear. Fig. 7.5 shows sections through safety footwear.

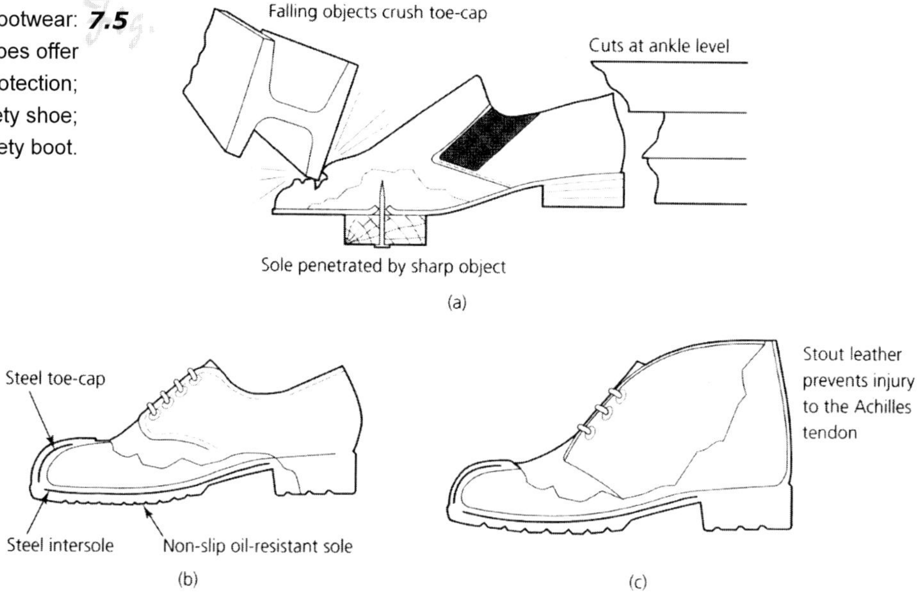

Footwear: **7.5**
(a) lightweight shoes offer no protection;
(b) industrial safety shoe;
(c) industrial safety boot.

7.7 LIFTING AND CARRYING

As shown by Fig. 7.1, the movement of materials is the biggest single cause of factory accidents. Manual handling accidents can be traced to one or more of the following:

- Incorrect lifting technique.
- Carrying too heavy a load.
- Incorrect gripping.
- Failure to wear protective clothing.

Fig. 7.6(a) shows the wrong technique for lifting, which can lead to ruptures, strained backs, sprains, slipped discs and other painful and permanent injuries. The correct technique is shown in Fig. 7.6(b). The back is kept straight and the lift comes from the powerful leg and thigh muscles. Fig. 7.6(c) is a reminder that, to avoid falls and injury, the load being carried must not obstruct forward vision.

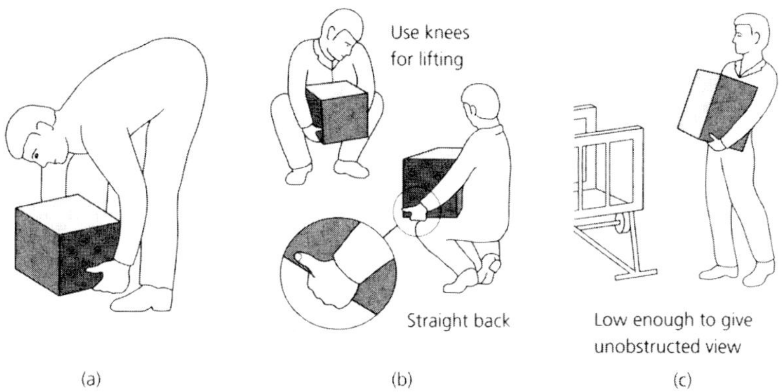

7.6 Lifting and carrying:
(a) an incorrect way to lift;
(b) the correct way to lift;
(c) the correct way to carry

Chapter 7

VOCABULARY

accident ; 사고, 재난
forethought ; 사전의 고려
premise ; 전제, 전제를 달다.
immeasurable ; 헤아릴 수 없는
suffer ; 경험하다, 겪다.
victim ; 희생
maim ; (손,발을) 불구로 만들다.
precaution ; 조심, 예방
protrude ; 내밀다, 불쑥 나오다.
paralyse ; 마비시키다.
liable ; 책임져야할, 책임이 있는
cuff ; 소매끝동
snatch ; 와락 붙잡다.
stout shoes ; 견고한 신발
distract ; 흩트리다.
abrasion ; 피부의 찰과상, 염증
barrier cream ; 보호크림
puncture ; (타이어)빵꾸내다. 찌름.
treading ; 밟다, 가다
crush ; 눌러 부수다, 뭉개다.
achilles' tendon ; 아킬레스건
scalp ; 머리가죽
extremely ; 극단적으로
infection ; (병 등의) 공기전염
environment ; 환경
entangle ; 뒤얽히게 하다
precaution ; 조심, 예방
inflammation ; 발화, 염증
abrasive ; 문질러 닳게 하는
grit ; (기계 등에 끼이는)잔모래
lid ; 뚜껑, 눈꺼플
exposure ; 드러남, 노출
ultraviolet ; 자외선의
radiation ; 방광, 방사, 방사물
intensity ; 강렬, 격렬
optic ; 눈의, 시력의
nerve ; 신경
splinter ; 쪼개진 조각
swarf ; (금속 등의)잘나낸 도막
toxic ; 유독한
inappropriate ; 부적당한
grime ; 때, 먼지
crucial ; 결정적인, 중대한
negotiate ; 교섭하다. 협정하다.
antiseptic ; 방부제의
water-soluble ; 물에 녹는
ingress ; 들어옴, 침입
discourage ; 용기를 잃게하다.
penetration ; 탄알의 관통
toecap ; (구두의) 콧등, 앞닫이
accordance ; 일치, 조화
rupture ; 파멸, 파괴
obstruct ; 차단하다, 장애물을 놓다.

8

Mechanical Fasteners

Chapter

(기계적인 체결부품)

8.1 INTRODUCTION

Mechanical fasteners with screw threads are the most common of all machine parts. It is important, therefore, that you understand how they are represented on drawings and how they are specified. You should also know the basic types of threads as well as their profiles.

8.2 SCREW THREAD TERMINOLOGY

The most widely used type of fastener is the screw thread configuration. Fig. 8.1 shows the design and terminology of a screw thread on the end of a bolt and nut. Such a thread is called an external thread, whereas the thread of a mating nut is an internal thread.

Chapter 8

Fig. **8.1** Terminology of a screw thread

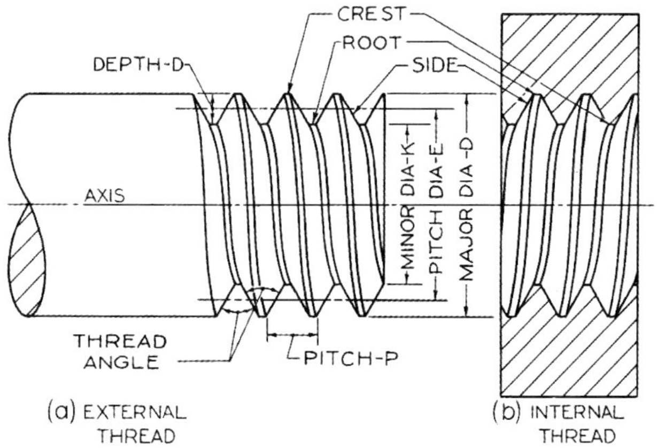

External thread. A thread on the outside of a member, as on a shaft.

Internal thread. A thread on the inside of a member, as in a hole.

Major diameter. The largest diameter of a screw thread (applies to both internal and external thread).

Minor diameter. The smallest diameter of a screw thread (applies to both internal and external threads).

Pitch. The distance from a point on a screw thread to a corresponding point on the next thread measured parallel to the axis. The pitch P is equal to 1 divided by the number of threads per inch.

Lead. The distance a screw thread advances axially in one turn.

Angle of thread. The angle included between the sides of the thread measured in plane through the axis of the screw.

Crest. The top surface joining the two sides of a thread.

Root. The bottom surface joining the sides of two adjacent threads.

8.3 SINGLE AND MULTIPLE THREAD

A **single** thread, as the name implies, is composed of one ridge, and the lead is therefore equal to the pitch. **Multiple** threads are composed of two or more ridges running side by side.

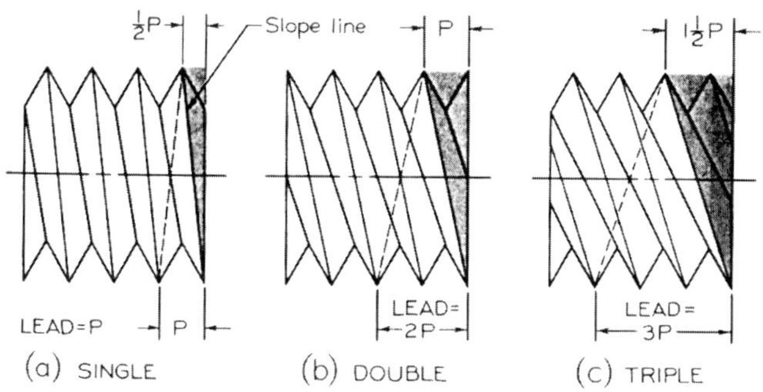

Fig. **8.2** Multiple thread

As shown in Fig. 8.2 (a)-(c), the **slope line** is the hypotenuse of a right triangle whose short side equals P/2 for single threads, P for double threads, 1 2P for triple threads, and so on. This applies to all forms of threads. In **double** threads, the lead is twice the pitch; in **triple** threads the lead is three times the pitch, and so on.

Multiple threads are used wherever quick motion, but not great power, is desired, as on fountain pens, toothpaste caps, valve stems, and the like.

8.4 SCREW-THREAD FORMS

Various forms of threads (Fig. 8.3) are used to hold parts together, to adjust parts with reference to each other, or to transmit power.

Screw thread forms *Fig.* **8.3**

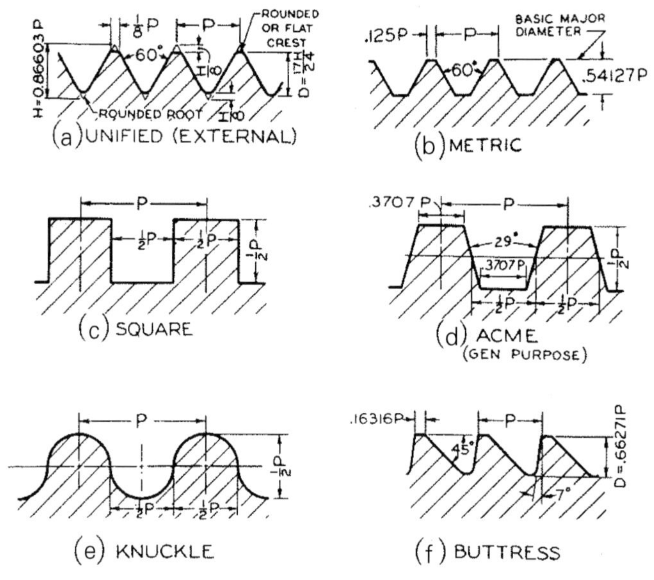

The **Unified thread** is the standard thread agreed upon by the United States, Canada, and Great Britain in 1948, and is gradually replacing the American National form. The crest of the external thread may be flat or rounded, and the root is rounded; otherwise, the thread form is essentially the same as the American National.

The **metric thread** is the standard screw thread agreed upon for international screw thread fasteners. The crest and root are flat, but the external thread is often rounded if formed by a rolling process. The form is similar to the American National and the Unified threads but with less depth of thread.

The **square thread** is theoretically the ideal thread for power transmission, since its face is nearly at right angles to the axis, but due to the difficulty of cutting it with dies and because of other inherent disadvantages, such as the fact that split nuts will not readily disengage, the square thread has been displaced to a large extent by the Acme thread. The square thread is not standardized.

The **Acme thread** is a modification of the square thread and has largely replaced it. It is stronger than the square thread, is easier to cut, and has the advantage of easy disengagement from a split nut, as on the lead screw of a lathe.

The **knuckle thread** is usually rolled from sheet metal but is sometimes cast, and is used in modified forms in electric bulbs and socket, bottle tops, and the like.

The **buttress thread** is designed to transmit power in one direction only and is used in large guns, in jacks, and in other mechanisms of similar high-strength requirements.

8.5 EXTERNAL THREAD SPECIFICATIONS

External threads are always specified in a note (Fig. 8.4). The note is made up of a seres of abbreviations. Thread information is always given in the same order: diameter of screw, number of threads per inch, kind(series) of thread, class of fit, and in the Unified series the letter A to designate external thread and the letter B to designate internal thread.

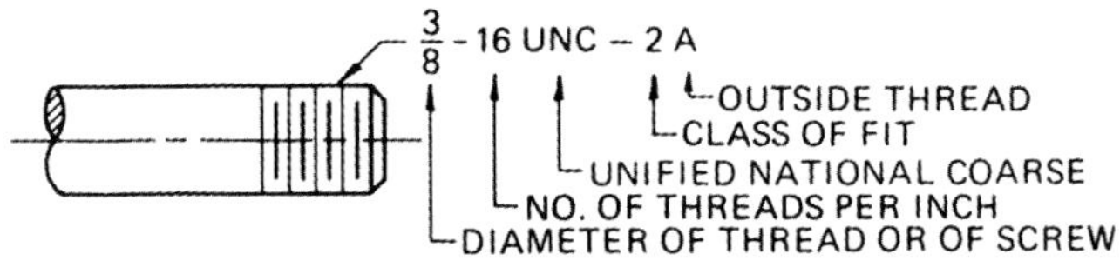

External thread **8.4** Fig.
specification, abbreviations, and translation

8.6 INTERNAL THREAD SPECIFICATIONS

Internal threads(Fig.8.5) are specified in the same general manner as external threads, and the specifications are given in the same order. Note the LH appears as the last item in the specifications. Threads are always assumed to be right-handed unless marked otherwise.

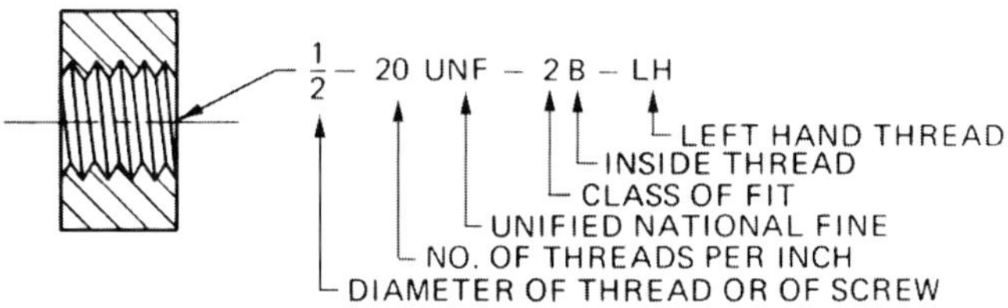

Fig. **8.5** Internal thread specification, abbreviations, translation.

8.7 COMMON SCREW FASTENERS

There are numerous types of common screw fasteners which can be utilized to connect one member to another. The following is a brief description of the more popular ones.

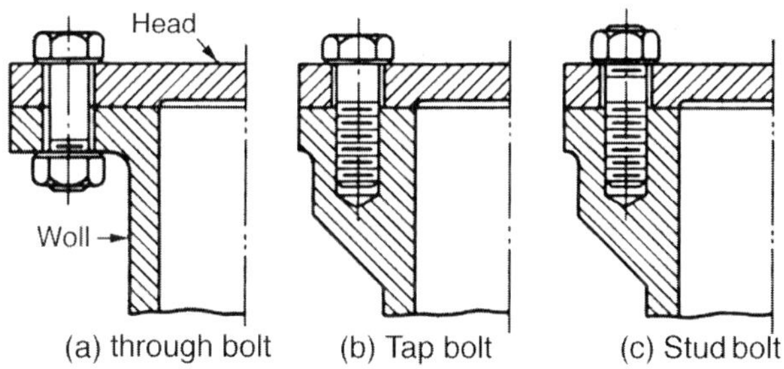

Fig. **8.6** Types of screw fastening

8.7.1 TROUGH BOLT

A through bolt has a head at one end and a nut fitted to the other, as shown in Fig. 8.6(a), where it used to attach a cylinder head to a cylinder.

8.7.2 TAP BOLT

Fig. 8.6(b) shows a tap bolt engaging in a tapped hole in the cylinder wall. If the cylinder head is removed frequently, the thread in the hole may become worn, necessitating costly repairs.

8.7.3 STUD BOLT

As shown in Fig. 8.6(c) stud bolt is threaded at both ends. One end is screwed into the tapped hole in the cylinder wall and the other end receives the nut.

8.7.4 EYEBOLT

An **eyebolt** (Fig. 8.7) is frequently used for providing a ring on a part for lifting purposes

Eyebolt *Fig.* **8.7**

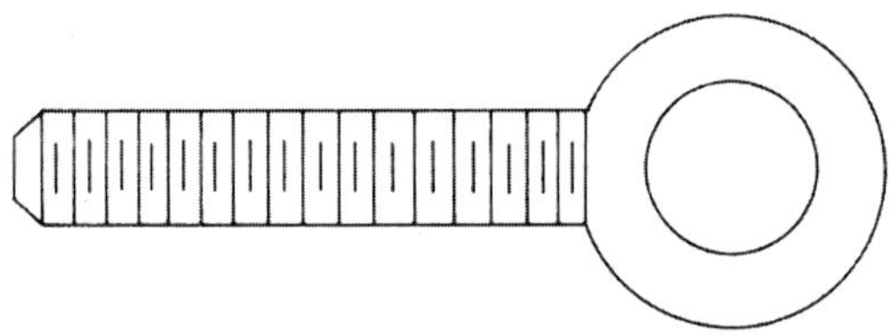

8.7.5 U-BOLT

A **U-bolt** is shown in Fig. 8.8. It uses two nuts for the fastening of parts. A popular application is the fastening together of the leaves of an automotive leaf spring.

U-Bolt *Fig.* **8.8**

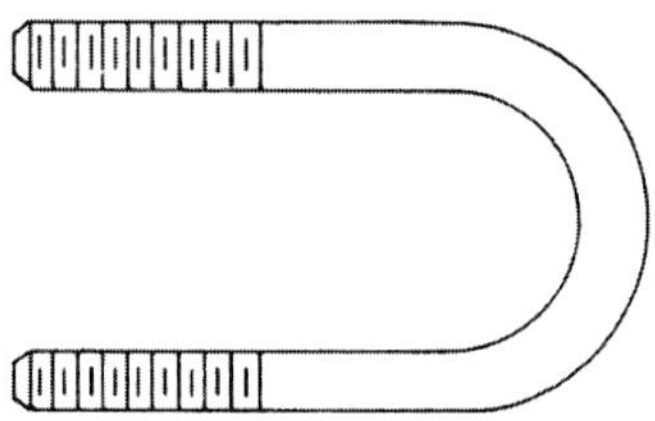

8.7.6 TURNBUCKLE

A **Turnbuckle** contains a central member which is threaded to two separate rods as illustrated in Fig. 8.9. One end of the central member contains a right-hand thread, whereas the other end contains a left-hand thread. As the central member is rotated, the rods are pulled closer together or pushed further apart depending on the direction of rotation of the central member. A typical application is for the tightening of cables.

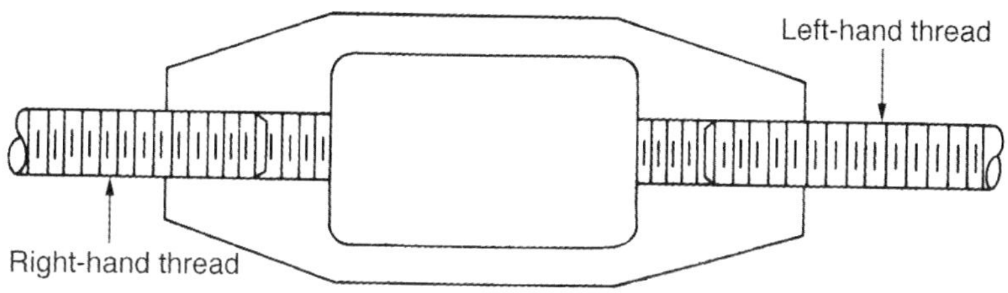

Fig. **8.9** Turnbuckle

8.7.7 CAP SCREW

A cap screw fastens a plate to another component as shown in Fig 8.10. When the parts are assembled, the cap screw passes through a clearance hole in one part and screws in a threaded hole in the other.

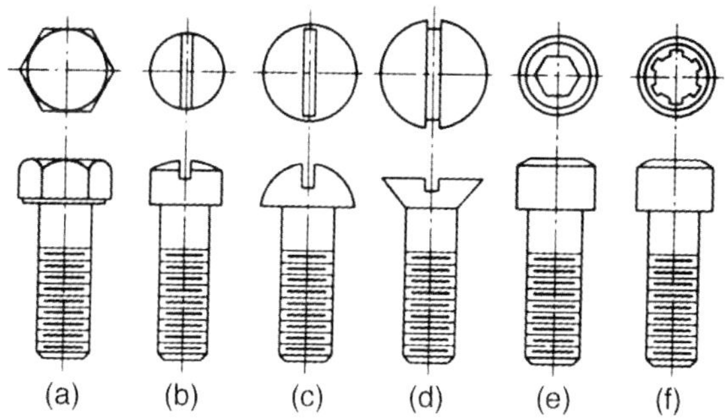

Fig. **8.10** Cap screw with various types of head; (a) hexagonal; (b) fillister; (c) button; (d) flat; (e) hexagonal socket; (f) fluted socket.

8.7.8 MACHINE SCREW

Machine screws are similar to cap screws, but are smaller in diameter. They are used mainly for Plate having thin sections as shown in Fig. 8.11.

Machine screw assembly Fig. 8.11

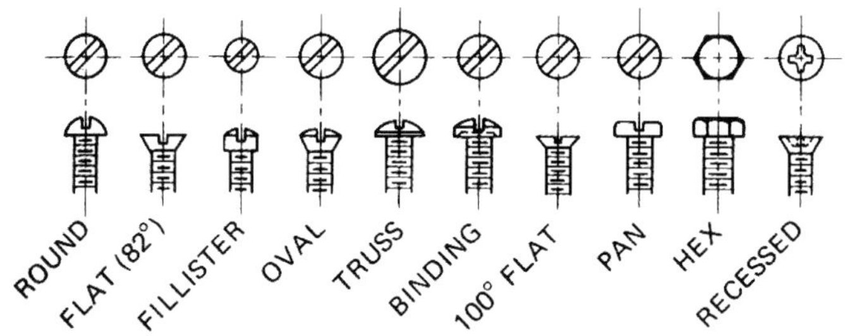

8.7.9 SETSCREWS

The **setscrew** is used to locate parts on a shaft and to prevent axial or rotary motion between the shaft and mounted component. A common application is the mounting of light-duty pulleys on shafts. Fig. 8.12 shows the various types of setscrews.

Types of setscrews Fig. 8.12

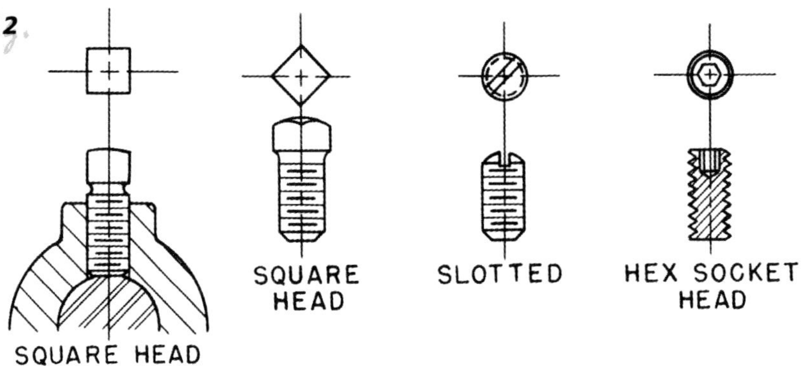

8.7.10 TAPPING SCREWS

Hardened tapping screws cut their own mating threads when driven into a hole of proper size (Fig. 8.13). They are superior to other fasteners in sheet-metal duct work because they require no nut. They are extensively used with plastics and nonferrous metals.

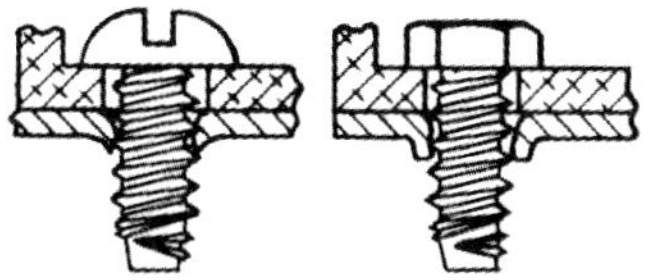

Fig. **8.13** Tapping screws

8.8 WASHERS

Three common types of washers are shown in Fig. 8.14. Plan washers are used to provide a smooth seat for the nut on castings and forgings when the holes are not spot faced. The chief function of split lock washers and spring lock washers is to prevent the nut from coming loose on the bolt.

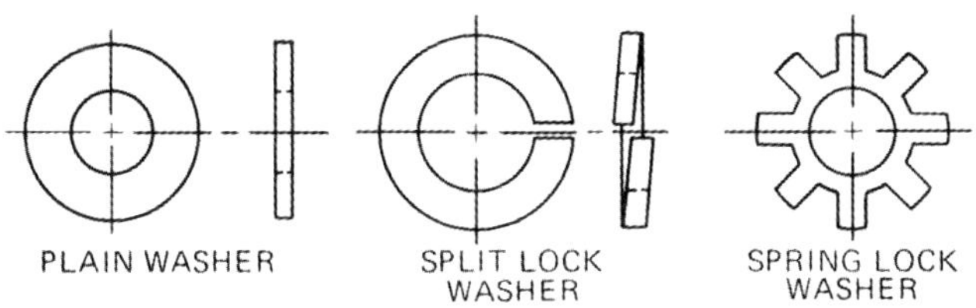

Fig. **8.14** Tree common types of washers

8.9 SNAP RINGS

Snap rings are spring-type fasteners which are used to secure components onto shafts or inside the bores of housings. There are two basic types as Follows

(1) External-type snap ring [Fig. 8.15(a)]

(2) internal-type snap ring [Fig. 8.15(b)]

Retaining ring Fig. **8.15**

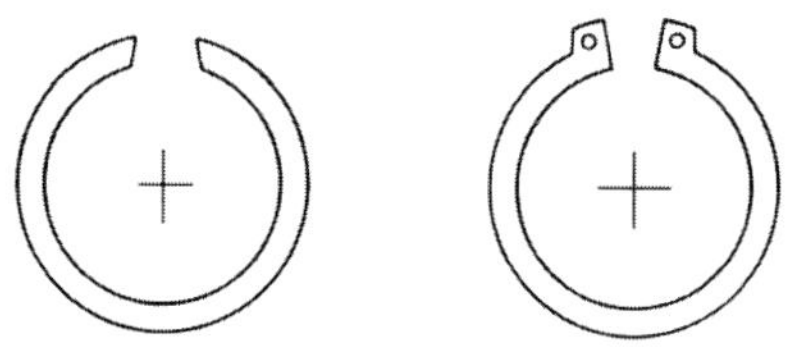

(a) External snap rings

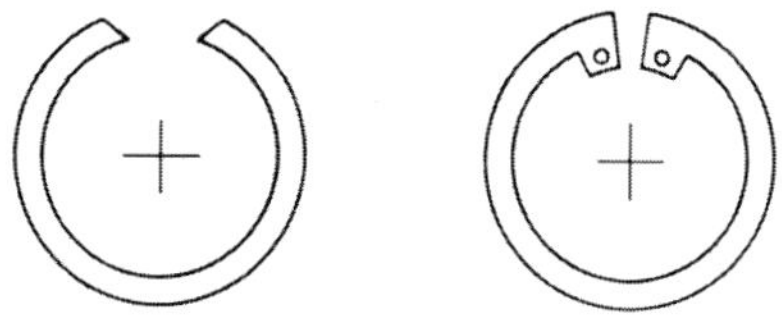

(b) Internal snap rings

Fig 8.16 shows snap rings used to retain a ball bearing in place. The larger ring supports an axial load on the shaft to the right in Figure. The smaller ring retains the inner race of the bearing against the shaft shoulder.

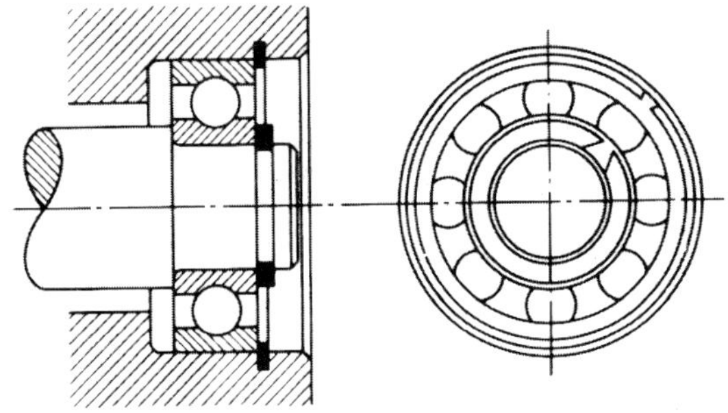

Fig. **8.16** ball-bearing installation.

8.10 KEYS

Keys are used to prevent relative movement between shafts and wheels, couplings, cranks, and similar machine parts attached to or supported by shafts, Fig. 8.17.

Square and **flat key** are suitable for heavy duty. A square key is shown in Fig. 8.17 (a) and a flat key in (b).

The widths of keys generally used are about one-fourth the shaft diameter. In either case, one-half the key is sunk into the shaft. The depth of the keyway or the keyseat is measured on the side-not the center, (a). Square and flat keys may have the top surface tapered $\frac{1}{8}''$ per foot, in which case they become square taper or flat taper keys.

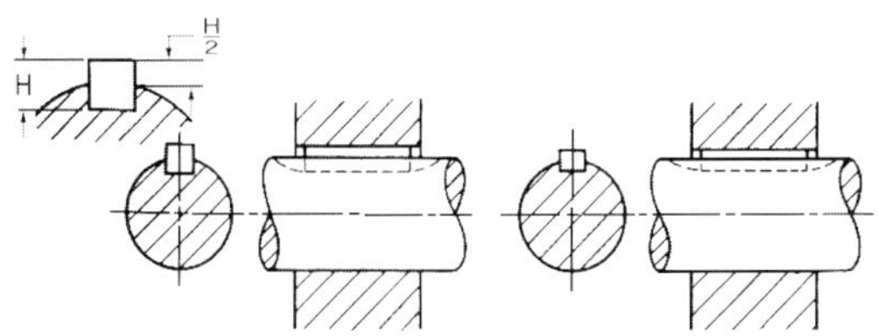

Fig. **8.17** Square and flat key

The **Woodruff key** is semicircular in shape, Fig. 8.18. This requires a key seat formed by a special side-milling cutter. This key will align itself in the key seat. It has the disadvantage of weakening the shaft more than by the use of a straight key.

Woodruff key *Fig.* **8.18**

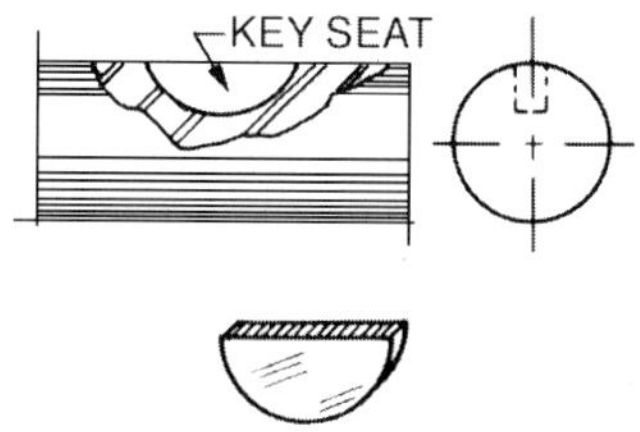

8.11 PINS

Pins include taper pins, straight pins, dowel pins, clevis pins, and cotter pins. For light work, the taper pin is effective for fastening hubs or collars to shafts, as shown in Fig. 8.19, in which the hole through the collar and shaft is drilled and reamed when the parts are assembled. For slightly heavier duty, the taper pin may be used parallel to the shaft as for square keys.

Taper pin *Fig.* **8.19**

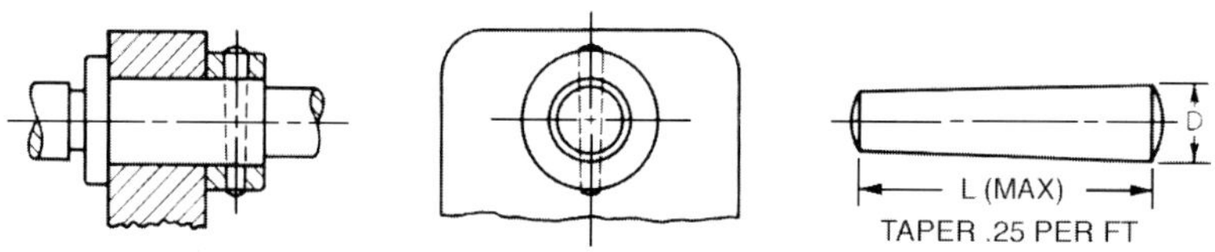

VOCABULARY

fastener ; 체결부품
profile ; 옆얼굴, 옆모습
configuration ; 형태, 윤곽
terminology ; 술어학, 술어
internal thread ; 암나사
correspond ; 일치하다. 부합하다.
adjacent ; 이웃의, 인접한
multiple thread ; 여러줄 나사
hypotenuse ; (직삼각형의) 사변
double thread ; 2줄나사
toothpaste cap ; 치약 뚜껑
unified thread ; 유니파이나사
acme thread ; 사다리꼴나사
knuckle thread ; 둥근나사
plain washer ; 평(平) 와셔
snap ring ; 멈춤링
axial load ; 축하중
inner race ; (구름베어링에서)내륜
installation ; 설치
woodruff key ; 반달키
utilize ; 이용하다. 소용되게 하다.
trough bolt ; 관통볼트
tap bolt ; 탭볼트, 삽입볼트
stud ; 박아 넣는 볼트
eyebolt ; 아이볼트, 고리볼트
U-bolt ; U자꼴의 볼트
leaf spring ; 판 스프링
turnbuckle ; 턴버클, 나사식 죔쇠
clearance ; 정리, 제거, 정돈
setscrew ; 고정나사
sheet metal ; 박판금(薄板金)
extensively ; 널리, 광범위하게
lock washer ; 로크워셔 (풀림방지용)
modification ; 조절, 변경
disengagement ; 자유, 해방, 이탈
split nut ; 분할너트
lead screw ; (선반) 어미나사
buttress thread ; 톱니나사
abbreviation ; 생략, 단축, 약어
designate ; 나타내다, 가리키다.

명언 / 격언

The government of the people, by the people, for the people, shall not perish from the earth.
November 19, 1863

— *Abraham Lincoln*

Fundamentals of Gears

9 Chapter

(기어의 기초)

9.1 INTRODUCTION

Gears are parts of interlocking toothed wheels that transmit force and motion in machines (Fig. 9.1). The four basic kinds of gears are rack and pinion, spur, bevel, and worm gears. In a pair of gear wheels, the smaller wheel turns more quickly than the larger one, and this difference in speed produces a difference in the force transmitted; the larger wheel turns with a greater force. Gears can therefore be used to increase or decrease a force, and to change the speed of a rotation, as well as its direction.

Fig. **9.1** Two meshing spur gears

Chapter 9

9.2 SPUR GEAR TERMINOLOGY

Spur gears are used to transmit rotary motion between parallel shafts; they are usually cylindrical, and the teeth are straight and parallel to the axis of rotation. The terms most frequently used in spur gearing are given in Fig. 9.2.

Fig. 9.2 Nomenclature for spur gear

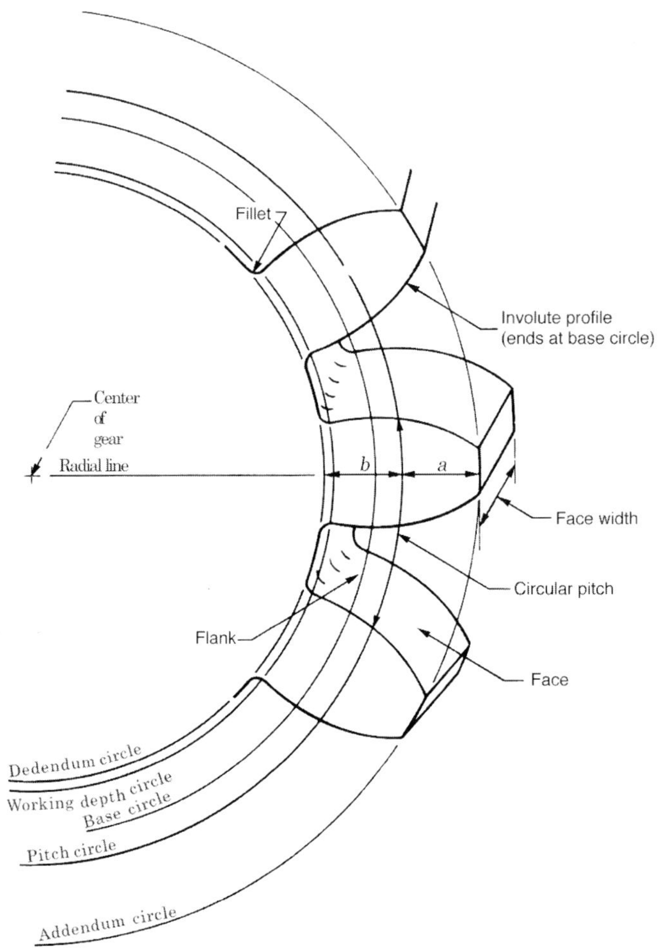

Pitch circle. The pitch circle is a theoretical circle on which all calculations are usually based. The pitch circles of a pair of mating gears are tangent to each other.

Pinion. A pinion is the smaller of two mating gears. The larger is often called the **gear** or the **wheel.**

Circular pitch. The circular pitch is the distance, measured on the pitch circle, from a point on one tooth to a corresponding point on an adjacent tooth.

Diametral pitch. The diametral pitch is the number of teeth on the gear per inch of pitch diameter. The units of diametral pitch are the reciprocal of inches. Note that the diametral pitch cannot actually be measured on the gear itself.

Module. The module is the ratio of the pitch diameter to the number of teeth. The customary unit of length is the millimeter. The module is the index of tooth size in SI, while the diametral pitch is used only with U.S. customary units.

Addendum. The addendum is the radial distance between the top land and the pitch circle.

Dedendum. The dedendum is the radial distance from the bottom land to the pitch circle.

Whole depth. The whole depth is the sum of the addendum and dedendum.

Clearance. The clearance is the amount by which the dedendum in a given gear exceeds the addendum of its mating gear.

Backlash. The backlash is the amount by which the width of a tooth space exceeds the thickness of the engaging tooth on the pitch circles.

9.3 RACK AND PINION

If the pinion in Fig. 9.3 is only allowed to move by revolving, the rack will be driven along in a straight line. Conversely if the rack is held and the pinion is free to move, rotation will make the pinion run along the rack in a straight line.

Action of rack and pinion **9.3** Fig.

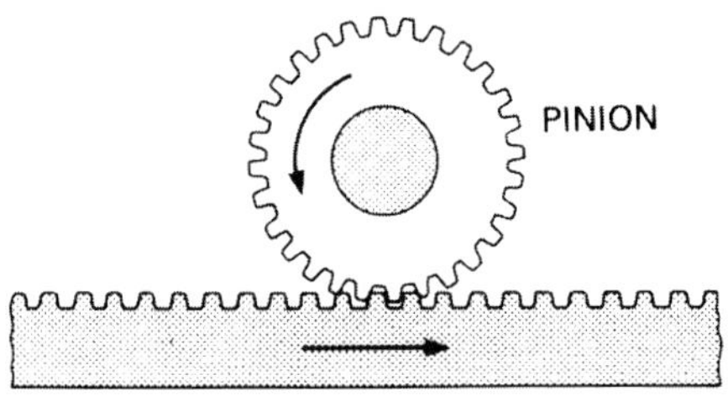

An application of this design is the feed movement of a drilling machine shown in Fig. 9.4. When the handle is rotated, the pinion attached to the handle shaft will drive the rack on the machine spindle in a straight line. This movement is used to feed the rotating drill into the work.

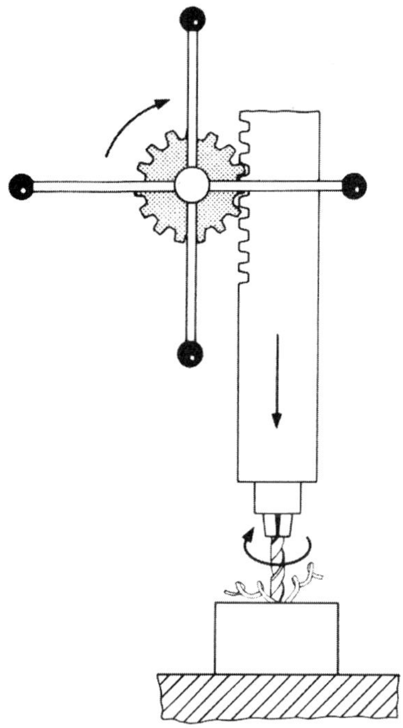

Fig. **9.4** Rack and pinion feed of a drilling machine

9.4 INTERNAL GEARS

All the gears we have discussed so far are external spur gears because their teeth are located on their outer surfaces. **Internal spur gears** (also called **annular gears**) are circular rings with teeth cut into the inner surfaces. Fig. 9.5 shows an internal gear and pinion. The pinion is usually the driver.

Internal spur gears with mating pinions provide much more compact drive systems than external spur gears, because the pinions lie within the internal gear. Since the pinion and internal gear rotate in the same direction, there is less sliding action and wearing of teeth as compared to external spur gears.

Internal gears Fig. 9.5

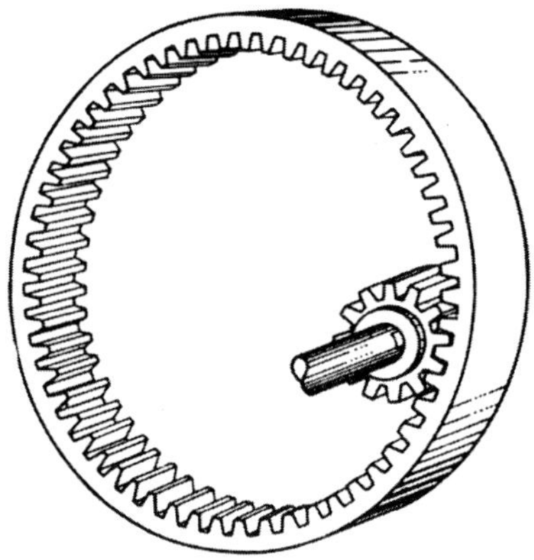

9.5 BEVEL GEARS

Bevel gears are characterized by their conical shape, as shown in Fig. 9.6 and 9.7. These are used for nonparallel shaft that would intersect if extended. The teeth are straight and are inclined to the gear axis by a conical angle called the pitch angle (Fig. 9.6 and 9.7).

The speed ratio of two meshing bevel gears is the ratio of the number of teeth in each gear. The speed ratio is also the ratio of their pitch diameters and the ratio of the sines of their pitch angles.

Fig. **9.6** Straight-tooth bevel gear

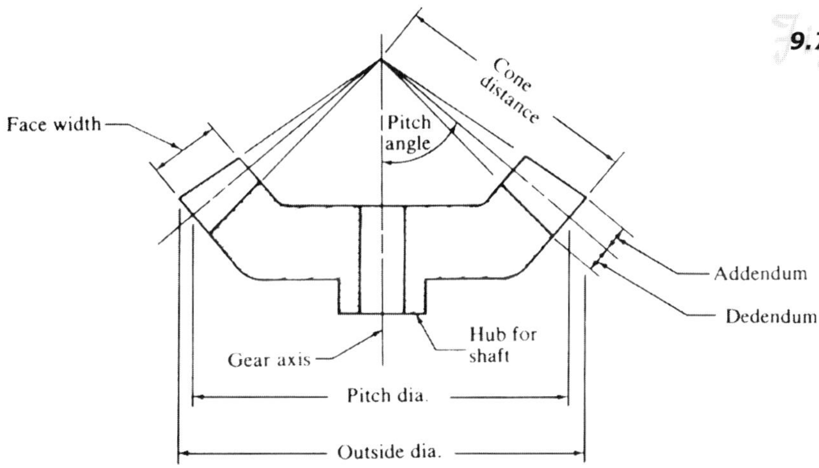

Fig. **9.7** Cross section of a bevel gear

Chapter 9

9.6 THE WORM AND WHEEL(WORMGEAR SET)

A **worm and wheel** are used for large speed reductions between two perpendicular but nonintersecting shafts (Fig. 9.8).

Worm and wheel ***Fig. 9.8***

The driver, which is called the **worm,** has a small diameter and low helix angle. In appearance it is similar to a threaded screw, and thus the helical teeth are commonly called **threads.** The follower is called the **worm gear** or **worm wheel** and has a face which is made concave to match the curvature of the worm. Hence, the wheel is said to **envelop** the worm, and their matching curvatures provide a large area of contact and reduce wear.

Worms can be made with either right- or left-hand threads similar to those used for threaded fasteners such as screws and bolts. Also, the worm can have single, double, triple, or quadruple threads.

9.7 GEAR TRAINS

A **gear train** is composed of two or more gears in mesh for the purpose of transmitting motion from one shaft to another. Ordinary gear trains are those in which the axes of none of the gears move relative to the frame and are of two types, simple gear trains and compound gear trains.

9.8 SIMPLE GEAR TRAINS

A **simple gear train** is one in which there is only one gear on each shaft such as in Fig. 9.9, where the gears are represented by their pitch circles. Gear A drives B, B drives C, C drives D, and D drives E. Let the numbers of teeth on the gears be N_A, N_B, etc. The ratio of the angular velocities of any pair of gears in mesh is the inverse of their number of teeth.

Thus

$$\frac{\omega_a}{\omega_B}=\frac{N_B}{N_A} \quad \frac{\omega_a}{\omega_B}=\frac{N_B}{N_A} \quad \frac{\omega_a}{\omega_B}=\frac{N_B}{N_A} \quad \frac{\omega_a}{\omega_B}=\frac{N_B}{N_A} \qquad (9.1)$$

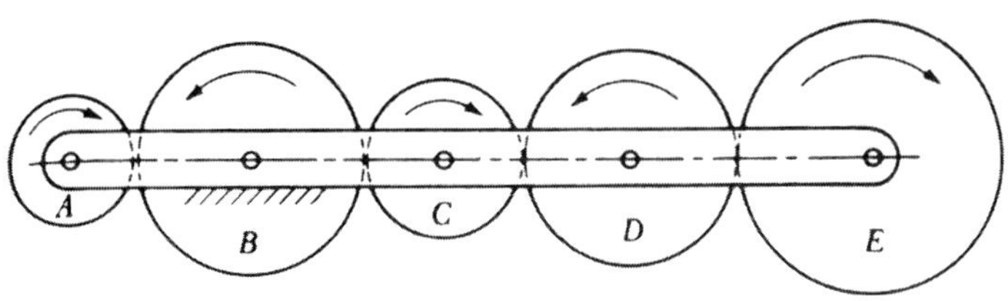

Fig. 9.9 Simple gear train

The **velocity ratio (VR)** of a gear train is the ratio of the angular velocity of the first gear in the train to the angular velocity of the last gear. For the train in Fig. 9.9 the velocity ratio is

$$VR=\frac{\omega_A}{\omega_E}=\frac{\omega_A}{\omega_B}\times\frac{\omega_B}{\omega_C}\times\frac{\omega_C}{\omega_D}\times\frac{\omega_D}{\omega_E}$$

and by substitution of Eqs. (9.1)

$$VR=\frac{N_B}{N_A}\times\frac{N_C}{N_B}\times\frac{N_D}{N_C}\times\frac{N_E}{N_D} \qquad (9.2)$$

The sign of the velocity ratio is considered positive if the first and last gears rotate in same direction and negative if they rotate in opposite directions. The easiest way to keep track of the directions of rotation is to place arrows on the gears. Later we will use signs to indicate directions of rotation. A plus sign will denote counterclockwise rotation, and a minus sign clockwise rotation.

From Eq. (9.2) we see that the velocity ratio for a simple gear train depends only on the number of teeth on last and first gears in the train because the tooth numbers on the intermediate gears cancel out. The latter are called **idler gears.** It is also obvious that the velocity ratio depends only on the numbers of teeth on the last and first gears because the pitch circles roll on one another and all the gears have the same pitch-line velocity.

Idler gears are used for two purposes: to connect gears where a large center distance is required, and to control the directional relationship between gears such as A and E in Fig. 9.9. Note that each idler added in the figure changes the direction of rotation of the last gear in the train.

9.9 COMPOUND GEAR TRAINS

A **compound gear train** is a gear train containing compound gears. In the compound gear train shown in Fig. 9.10, the numbers indicate the number of teeth on each gear.

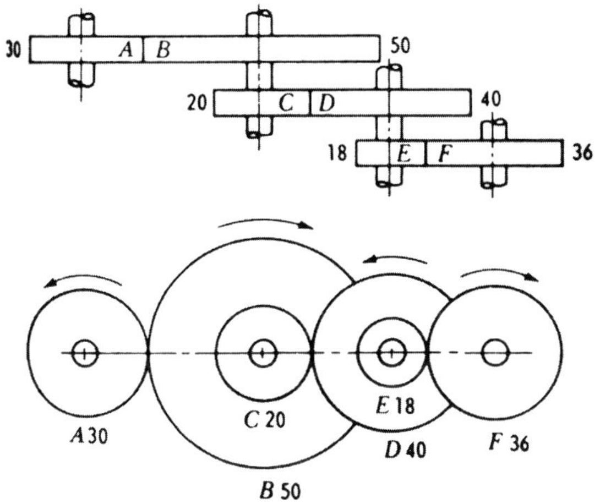

Fig. **9.10** Compound gear train

If the speed of A is 1600 r/min ccw, the speeds of the various gears will be

$$\omega_A = +1\,600 \text{ r/min}$$

$$\omega_B = \omega_C = \frac{30}{50} \times 1\,600 = -960 \text{ r/min}$$

$$\omega_B = \omega_C = \frac{20}{40} \times 960 = 480 \text{ r/min}$$

$$\omega_F = \frac{18}{30} \times 480 = -240 \text{ r/min}$$

The velocity ratio is

$$VR= \frac{\omega_A}{\omega_F} = -\frac{600}{240} = -\frac{20}{3}$$

For a compound gear train the velocity ratio can be written

$$VR = \frac{\text{product ofteeth on driven gears}}{\text{product of teeth on driving gears}}$$

$$= \frac{50}{3} \times \frac{40}{20} \times \frac{36}{18} = - \frac{20}{3}$$

The advantage of a compound train over a simple gear train is that a much larger speed reduction from the first to the last shaft can be obtained with small gears. If a simple gear train were used to give a large speed reduction, the last gear would have to be large. Usually for a speed reduction in excess of 7 to 1 a simple train is not used; instead a compound train or worm gearing is employed.

VOCABULARY

fundamentals ; 기초, 토대
interlock ; 연결하다, 결합하다
spur gear ; 스퍼 기어
cylindrical ; 원통형(圓筒形)의
axes ; axis(축)의 복수형
theoretical ; 이론(상)의
calculation ; 계산, 계산의 결과
tangent ; 탄젠트, 접하는
pinion ; 작은 톱니바퀴
correspond ; 일치하다, 부합하다.
diametral pitch ; 지름피치
reciprocal 상호간의, 역수
module ; 모듀울, 기어에서 치(齒)의 크기를 나타내는 용어
customary unit ; 관습적인 단위
addendum ; 이끝높이
clearance ; 간극
backlash ; 뒤틈
rack ; 랙 (피치원 지름이 무한대인 기어)
internal gear ; 내치차
annular gear ; 링기어
bevel gear ; 베벨기어, 베벨휘일
characterize ; 특징 지우다,
nonparallel ; 평행하지 않은
intersect ; (선 또는 면) 교차하다
perpendicular ; 수직의
appearance ; 출현, 외간
envelop ; 싸다, 봉하다.
concave ; 오목한 , 오목렌즈
curvature ; 안곡, 뒤틀림
guardruple thread ; 4줄나사
gear train ; 기어열
idler gear ; 유차
intermediate ; 중간의

명언 / 격언

There is a great deal of difference between the eager man who wants to read a book, and the tired man who wants a book to read.

– G. K. Chesterton

Chapter 9

Thomas Alva Edison

(1847 ~ 1931)

Thomas Alva Edison. Considered by many as one of the greatest inventors in history, he was born in Milan, Ohio. He obtained patents in such fields as telegraphy, phonography, electric lighting and photography. In 1882, he designed the first hydroelectric plant in Appleton, Wisconsin. In 1887, he and Sir Joseph Wilson Swan produced the Edison electric lamps. As a boy, Edison had only three months of formal schooling. He was taught at home by his mother, a former teacher. He changed the lives of millions of people with such inventions as the electric light and the phonograph. In his lifetime, he patented 1,093 inventions. After the death of his first wife, he married again. He had six children, three by each wife. He valued long, hard work. One of his famous sayings was;

"Genius is 1 percent inspiration and 99 percent perspiration."

10 Chapter

Strength of Materials

(재료역학)

10.1 STRESS

When a material has a force exerted on it, the material is said to be **stressed** or in a state of **stress.** If a rod is subjected to a **tension,** the force per unit area of cross-section of the rod is referred to as **tensile stress.**

If the rod is subjected to compression, the force per unit area of cross-section is termed a **compressive stress.**

Hence, in general:

$$\text{stress} = \frac{\text{force}}{\text{cross sectional area}}$$

10.2 STRAIN

When a rod or a wire is pulled, it stretches; and the total stretch or elongation, expressed as a fraction of the unstretched length, is termed **direct strain** or merely **strain** when it is obvious that the change in length is due to tension or compression.

Thus, if the pull on a rod having an unstretched length of 1 m produces an extension of 2.5 mm,

$$\text{strain} = \frac{\text{extension}}{\text{original length}} = \frac{2.5}{1 \times 1000} = 0.0025$$

10.3 HOOK' S LAW

Robert Hook (1635-1703), a famous English physicist, architect and inventor, carried out a series of experiments with springs and wires and was the first to publish, in 1676, a clear statement to the effect that when a material is worked within its elastic range, **the extension is proportional to the force.**

10.4 MODULUS OF ELASTICITY OR YOUNG' S MODULUS

For a material worked within its elastic range, we have from Hook' s law that since the extension is proportional to the force:}}

$$\text{strain} \propto \text{stress},$$

$$\text{i.e.} \quad \frac{\text{stress}}{\text{strain}} = \text{a constant}.$$

This constant is termed **modulus of elasticity** or **Young' s modulus**, after Thomas Young* who was the first to determine this constant for some materials. The symbol for modulus of elasticity or Young' s modulus is E,

$$\text{i.e.} \quad E = \frac{\text{tensile or compressive stress}}{\text{strain}}$$

$$= \frac{\text{force/originalcross-sectional area}}{\text{change of length/original length}}$$

* *Thomas Young (1773~1829) was one of the most versatile geniuses in history. He studied languages and medicine and was a Professor of Physics at the Royal Institution, London. He was the first to translate the Egyptian hieroglyphics on the Rosetta Stone at the British Museum and was also famous for his share in establishing the undulatory theory of light.*

10.5 STRESS-STRAIN GRAPH FOR MILD STEEL

Fig. 10.1 shows a typical graph for a tensile test on a mild-steel specimen. The stresses have been calculated by dividing the loads by the **original** cross-sectional area of the test piece, and the strains by dividing the extensions by the **original** length selected for the test.

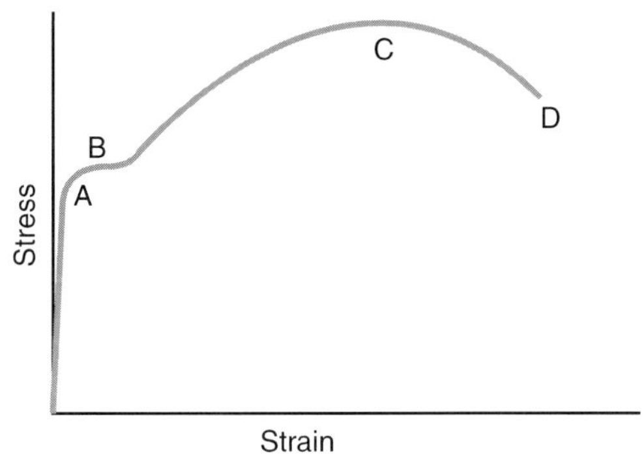

10.1 Stress-strain graph for mild steel

Chapter 10

Point A in Fig. 10.1 marks approximately the **limit of proportionality** of stress to strain, and is practically the **elastic limit**, i.e. the highest stress that can be applied without producing permanent deformation. For stresses up to the elastic limit, the material returns to its original length on removal of the load.

Portion B in Fig. 10.1 gives the **yield stress**, namely the stress at which, in a tensile test, elongation of the test piece first occurs without increase of load. This plastic yielding is not found in all ductile materials but is a marked feature of the softer irons and steels.

The maximum stress, represented by C in Fig. 10.1 is termed the **tensile strength**,* and is obtained by dividing the maximum load applied during a tensile test by the **original** cross-sectional area of the specimen and not by the reduced area of section after plastic extension. At about point C, the cross-sectional area at some point along the length of the test specimen begins to decrease rapidly and the load required to increase the extension beyond C consequently decreases until the specimen ultimately fractures at a load corresponding to point D.

* *This term replaces ultimate tensile stress, the use of which is not now recommended by the British Standards Institution.*

10.6 TENSILE STRENGTH AND FACTOR OF SAFETY

The tensile strength of a material has already been referred to as the maximum tensile load (in a test) divided by the original cross-sectional area of the specimen. It is obvious that no such stress must be allowed in any part of a machine or structure. The **permissible stress** should, in fact, always be far below the tensile strength and also almost always below the stress corresponding to the elastic limit. The ratio of the tensile strength to the permissible tensile stress is termed the **factor of safety.**

i.e.

$$\text{factor of safety} = \frac{\text{tensile strength}}{\text{permissible tensile stress}}$$

Hence,

$$\text{permissible tensile stress} = \frac{\text{tensile strength}}{\text{factor of safety}}$$

The factor of safety to be adopted depends upon many circumstances and in often covers many contingencies very imperfectly known.

The following table gives approximate values of the tensile strength and Young' s modulus for several of the most commonly used materials, but materials having a particular name vary widely in their properties and often more precise information is required about the actual materials to be employed. This is particularly true in regard to cast irons and steels of widely differing kinds.

TENSILE STRENGTH AND YOUNG' S MODULUS

Material	Tensile strength in MPa	Young' s modulus in GPa
Wrought iron	300 to 400	190
Mild steel	450 to 600	210
Cast iron	120 to 160	120
Copper(hard drawn)	300 to 350	110
Aluminium(hard drawn)	150 to 190	70

10.7 SHEARING

A material is under the action of shear **stress across** a surface within it when there are forces tending to make the part on one side of the surface slide past the part on the other side of the surface. Thus a pin or rivet shown in Fig. 10.2 is subject to shear stress across the plane surface (dotted) AB. The upper A of the pin tends to slide to the left relative to the lower part B.

Shear stress. ***Fig.* 10.2**

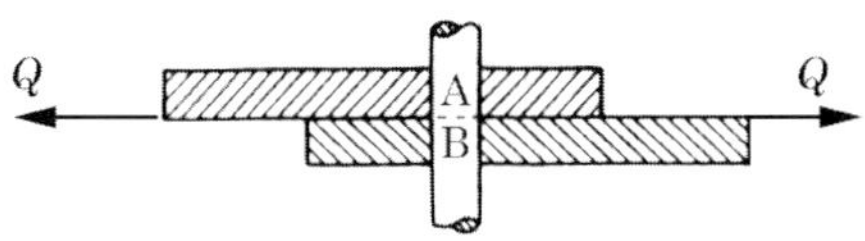

The tendency of the stress is to **shear** the pin into two parts, A and B. When shears are used for cutting metal or vegetation, actual shearing fracture takes place; but in parts of machines and structures, the **shear stress** is much below the **shear strength**. The average shear stress in Fig. 10.2 is :

$$\frac{\text{shear force}}{\text{area of cross-section between A and B}}$$

A common case of the use of metal to resist shearing is to be found in riveted joints and in instances where one part of a machine is capable of turning relative to another about a pin. For example, in the knuckle joint shown in Fig. 10.3, the pin resists the pull at **two** sections, A and B, so that the average shear stress in the pin is equal to force Q divided by **twice** the area of cross-section of the pin. The pin is said to be subjected to **double shear.**

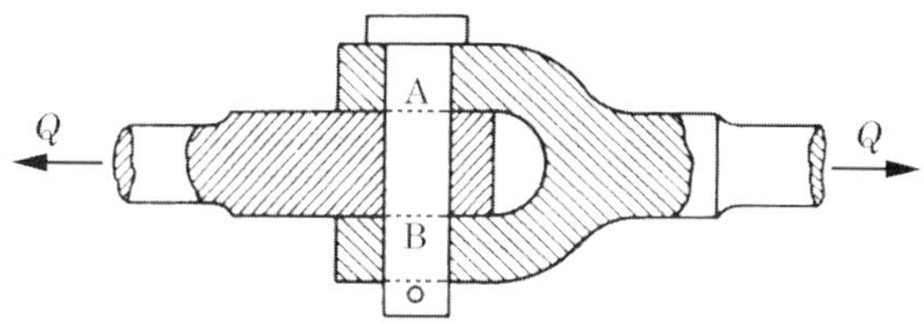

10.3 Pin in double shear.

Chapter 10

VOCABULARY

stress ; 응력
exert ; 힘 등을 쓰다, 움직이다.
tensile stress ; 인장력
compressive stress ; 압축응력
strain ; 잡아당기다
elongation ; 신장
fraction ; 파편, 단편
physicist ; 물리학자
architect ; 건축가
inventor ; 발명가
publish ; 발표하다, 출판하다
elastic range ; 탄성범위
proportional ; 비례하는
modulus ; 율, 계수
elasticity ; 신축성
young' s modulus ; 영계수
determine ; 결정하다
extension ; 신장, 뻗음
elastic-limit ; 탄성한도
permanent ; 영속하는, 영구적인
deformation ; 변형
yield stress ; 항복응력
ductile material ; 연성재료
tensile strength ; 인장강도
reduce ; 줄이다, 감소하다.
consequently ; 따라서, 그 결과로
ultimately ; 최후로, 마침내
permissible stress ; 허용응력
circumstances ; 사정, 상황
contingency ; 우발사건
imperfectly ; 불충분하게
approximate ; 접근하다, 대략
shear stress ; 전단응력
vegetation ; 초목, 식물
riveted joint ; 리벳이음
knuckle joint ; 너클 조인트
double shear ; 이중전단

명언 / 격언

First impressions are most lasting.

— *Dickens*

11

Fluid Mechanics Chapter

(유체역학)

11.1 INTRODUCTION

A **fluid** is a **gas** or **liquid** that conforms to the shape of its container. In a car, antifreeze mixtures, brake fluid, lubricating oils and gases in cylinders are examples of fluids. A **hydraulic system** is a fluid system that uses a liquid as the fluid. A **pneumatic system** is a fluid system that uses air or gas as the fluid.

Air-conditioning systems use fluids to control temperature and humidity in homes, offices and factories. They operate by using fans to create pressure in ductwork that circulates air through buildings.

City water systems are important fluid systems (Fig. 11.1) made up of elevated water tanks, underground pipes, Water meters, house plumbing and faucets. In air-conditioning systems, the fluid is air. In city water systems, the fluid is water.

Chapter 11

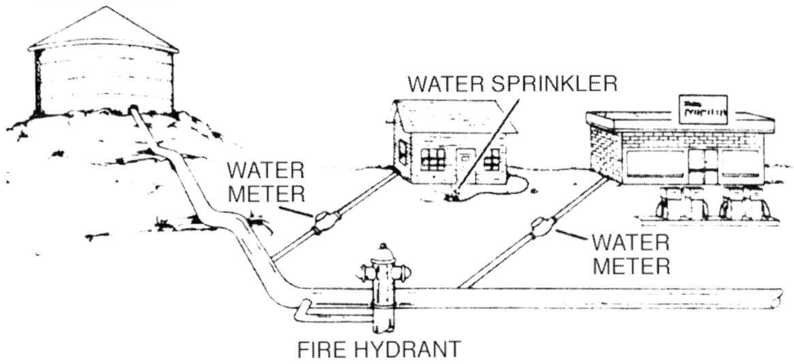

Fig. 11.1 A city water system depends on pressure to move water

11.2 DENSITY AND SPECIFIC GRAVITY

Have you ever wondered why motor oil floats on water? Or why a balloon filled with hot air rises? The answer is found in a property of materials called **"density"**. The density of a substance is the mass of that substance divided by the volume of that substance. See the following equation:

$$\text{Density} = \frac{\text{Mass}}{\text{Volume}}$$

$$D = \frac{M}{V}$$

The **specific gravity** of a substance is the density of that substance divided by the density of water. The following equation explains this relationship:

$$\text{Specific Gravity} = \frac{\text{Density of Substance}}{\text{Density of Water}}$$

Since specific gravity is **"density divided by density,"** units cancel out. Therefore, specific gravity is always a pure number. For example, the specific gravity of mercury is 13.6.

Table 1.3 lists the density and specific gravity of some common solids and liquids. Note that density always requires units-like gm/cm^3. Specific gravity has the same numerical value as density, for any given substance, but carries no units. For example, copper has a **density** of **8.9 gm/cm^3**, and a **specific gravity** of **8.9.**

TABLE 1.3 DENSITY AND SPECIFIC GRAVITY

	Density	Specific Gravity
Solids		
Gold	19.3 gm/cm^3	19.3
Lead	11.3 gm/cm^3	11.3
Silver	10.5 gm/cm^3	10.5
Copper	8.9 gm/cm^3	8.9
Steel	7.8 gm/cm^3	7.8
Aluminum	2.7 gm/cm^3	2.7
Balsa Wood	0.3 gm/cm^3	0.3
Oak Wood	0.8 gm/cm^3	0.8
Liquid		
Mercury	13.6 gm/cm^3	13.6
Water	1.0 gm/cm^3	1.0
Oil	0.9 gm/cm^3	0.9
Alcohol	0.8 gm/cm^3	0.8
Ethyene Glycol	1.125 gm/cm^3(32° F)	1.125
(Antifreeze)	1.098 gm/cm^3(77° F)	1.098

11.3 BUOYANT FORCE

When an object floats or sinks in a fluid, there's an upward force on the object that's equal to the weight of the fluid displaced. This upward force is called **"buoyant force"**. The fluid displaced is the amount of fluid "pushed out of the way" by the object in the fluid. If the density of the object is greater than the density of the fluid, its weight is geater than the buoyant force. So it sinks. For example, iron sinks in water.

If the object is less dense than the fluid in which it's placed, its weight is less than the buoyant force. So the object floats. For example, balsa wood floats easily in water. Use Table 1.3 to help you answer this question: "Will lead sink or float in mercury?"

Fig. 11.2 shows an object floating in a liquid. The weight of the liquid displaced by the object equals the weight of the object. If the weight of the object increases, it floats lower in the liquid, just as a boat floats lower when loaded.

Buoyant force Fig. 11.2

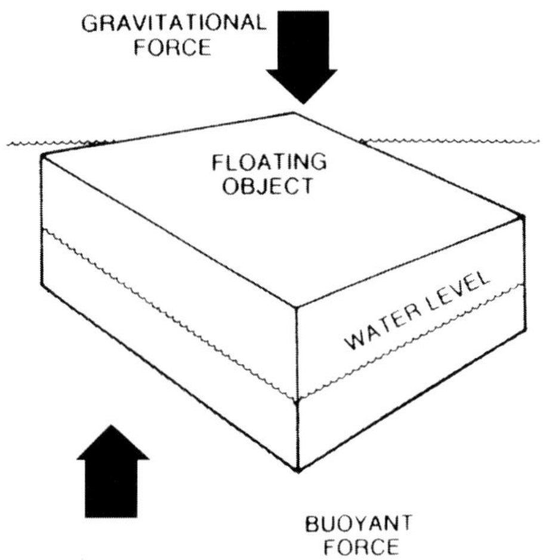

11.4 PRESSURE

When the tire of a bicycle is inflated, the compressed air exerts an outward thrust on the inner surface of the tire, and the thrust per unit area is termed the **pressure.** The SI unit of pressure is **the newton per square metre or pascal,** exactly the same as for stress. The atmospheric pressure known as the **standard atmospheric pressure** is 101.325 Pa or N/m^2.

The term **pressure** generally refers to the thrust per unit area exerted by **fluids** such as water, air, steam, etc., whereas the term **stress** refers to the force per unit area exerted in a solid body. Sometimes the term "pressure" is used to denote the total thrust on a surface; but to avoid confusion, it is better in such a case to use the term "thrust" or " force."

11.5 PRESSURE INCREASES WITH DEPTH

Suppose the tank in Fig. 11.3 to be nearly filled with a liquid having a density of ρ kilograms per cubic meter. Let us consider a

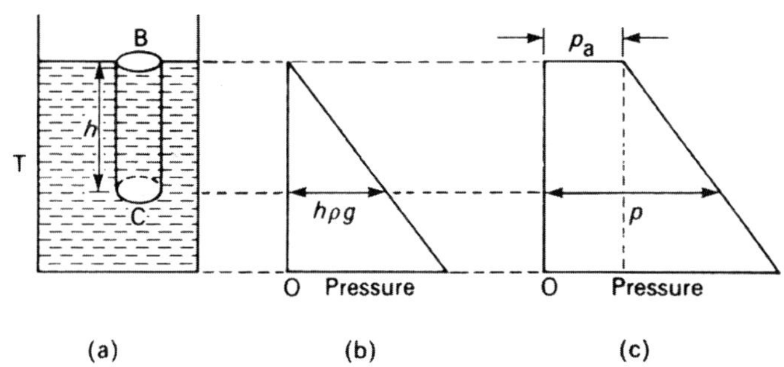

Fig. **11.3** Variation of pressure with depth in a liquid.

vertical cylindrical column BC of hight h meters and cross sectional area α square meters.

If the upper surface B is on a level with the surface of the liquid, and if the effect of atmospheric pressure on the surface of B is neglected, the only vertical force acting downward on the bottom surface at C is the **weight** of the liquid enclosed in cylinder BC.

Volume of liquid in cylinder BC = αh cubic meters,

∴ mass of liquid in cylinder BC = $\alpha h \rho g$ kilograms.

If g is the gravitational force, in newtons, on a mass of 1 kg,

weight of liquid in cylinder BC = $\alpha h \rho g$ newtons

and

pressure on lower surface of BC = $\alpha h \rho g / a = h \rho g$ pascals

Since the liquid is assumed stationary, the downward pressure at C is balanced by an equal upward pressure. This equilibrium applies to every pair of forces acting in opposite directions throughout the liquid. Hence, if the atmospheric pressure on the surface of the liquid is neglected, **the pressure is directly proportional to the depth**, as shown in Fig. 11.3(b).

If the atmospheric pressure on the surface of the liquid is ρ_a, then the total pressure p at depth h below the surface is given by:

$$p = \rho_a + h\rho g$$

as shown in Fig. 11.3(c).

11.6 PASCAL'S PRINCIPLE

The pressure applied to an enclosed fluid is transmitted undiminished to every point in a fluid and to the walls of the container.

This happens almost instantaneously for an incompressible fluid; in a gas it takes some time for equilibrium to be reached.

11.7 HYDRAULIC PRESS

Consider the **hydraulic press** shown in Fig. 11.4. The force applied to piston **A** creates a pressure of $P = F_1/A_2$ in the liquid.

This pressure is transmitted to every part of the liquid, including the liquid in cylinder B. This pressure will exert a force F_2 on piston **B** that is equal to $\rho \times A_2$. The relationship between F_1 and F_2 can be established as:

$$\frac{F_1}{A_1} = P = \frac{F_2}{A_2}$$

By selecting cylinder sizes so that $A_2 > A_1$, the system can be used to multiply force.

Fig. 11.4 Cutaway view of the cylinders and pistons in a hydraulic press

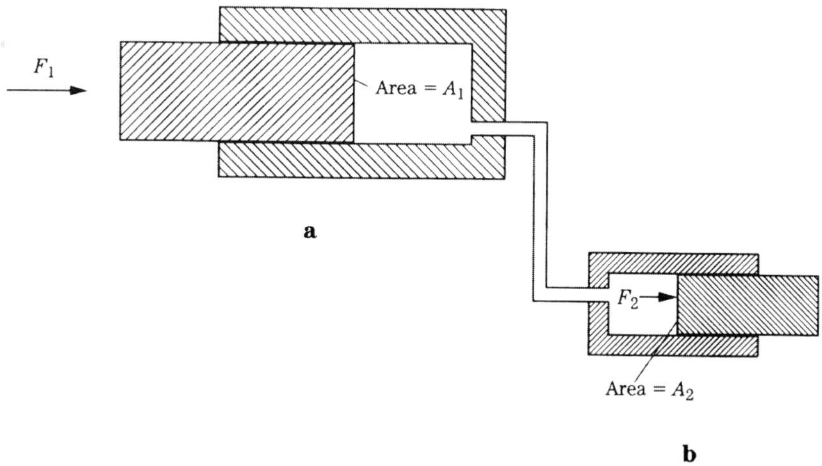

11.8 HYDRAULIC LIFT

A hydraulic lift at the local service station is used to lift cars and trucks off the ground. It works because liquids are incompressible (liquid can' t be squeezed into a smaller volume), and liquids transmit pressure equally in all directions over a long distance.

The air compressor shown in Fig. 11.5 increases the pressure above the hydraulic fluid. This pressure is then transmitted to the bottom surface of the lifting "piston." Because the pressure in the fluid is high, and the area of the lifting piston is large, a large pushing force is exerted on the lifting piston. This force enough to lift the truck.

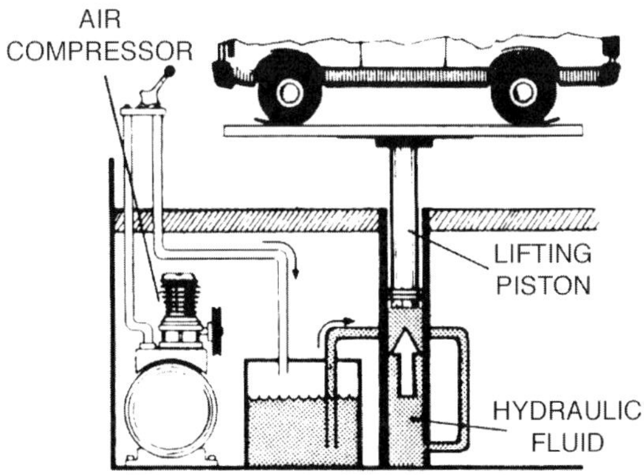

Fig. **11.5** A hydraulic lift

Chapter 11

VOCABULARY

fluid system ; 유체장치
liquid ; 액체의, 투명한
hydraulic system ; 유압장치
pneumatic system ; 공압장치
humidity ; 습기, 습도
temperature ; 온도
ductwork ; 공기조절(장치)
circulate ; (공기등이) 순환하다
underground pipes ; 지하매설관
plumbing ; , (수도 가스의) 배관공사
equilibrium ; 평형, 균형,
undiminished ; 줄지 않는
instantaneously ; 순간적으로,
incompressible ; 비압축성의
hydraulic lift ; 수압 승강기
service station ; 정비공장
buoyant force ; 부력
balsa wood ; 벌사나무(아메리칸 산 관목의 일종, 가볍고 단단함)
tyre ; 타이어(=tire)
inflate ; 팽창하다.
atmospheric pressure ; 대기압
denote ; 표시하다, 나타나다.
confusion ; 혼동, 혼란
cubic meter ; 입방미터
gravitational force ; 중력사
faucet ; (수도 등의) 물꼭지
density ; 밀도, 농도
specific gravity ; 비중
float ; (물 위에) 뜨다.
balloon ; 기구, 풍선
mercury ; 수은

명언 / 격언

A good book is the best of friends, the same today and forever.

— *English Proverb*

Archimedes and the Crown
(287BC~212BC)

"EUREKA!"

Archimedes (287 BC~212 BC) is said to have discovered his Principle on seeing the water rise as he got into his bath.

King Hieron had asked him to find out whether the King' s new crown was pure gold, without damaging it. At the palace he lowered the crown into a jug of water and noted how far the water rose. He then did the same with a pure gold of the same weight and the water rose by less, proving that the crown contained another, less dense, metal.

12 Chapter

Internal Combustion Engine (내연기관)

12.1 INTRODUCTION

A steam engine works in two stages: a fire in a boiler produces steam, and then this expands in a cylinder to do the work. By the mid-19th century people were experimenting with smaller, more efficient engines that did away with steam by putting the fire inside the cylinder itself. There were problems finding a suitable fuel, getting it into the engine, and setting fire to it.

These were all solved by the German engineer Nicklaus Otto (1832-1891). He built the first petrol engine in 1861 and followed this up with the "four stroke" engine, the ancestor of the modern car engine, in 1876. Otto used an electric spark to ignite the mixture of fuel and air.

In 1893 another German engineer, Rudolph Diesel (1858-1913), produced an engine in which the mixture is made to explode simply by being compressed. Diesel engines are heavier, but more reliable and economical than petrol engines.

Chapter 12

12.2 FOUR-STROKE CYCLE ENGINE

The diagrams below show how a single-cylinder petrol engine works. The engine is called a four-stroke engine because it keeps repeating a series of four up-and-down movements or strokes (Fig. 12.1).

Four-stroke cycle *Fig.* **12.1**

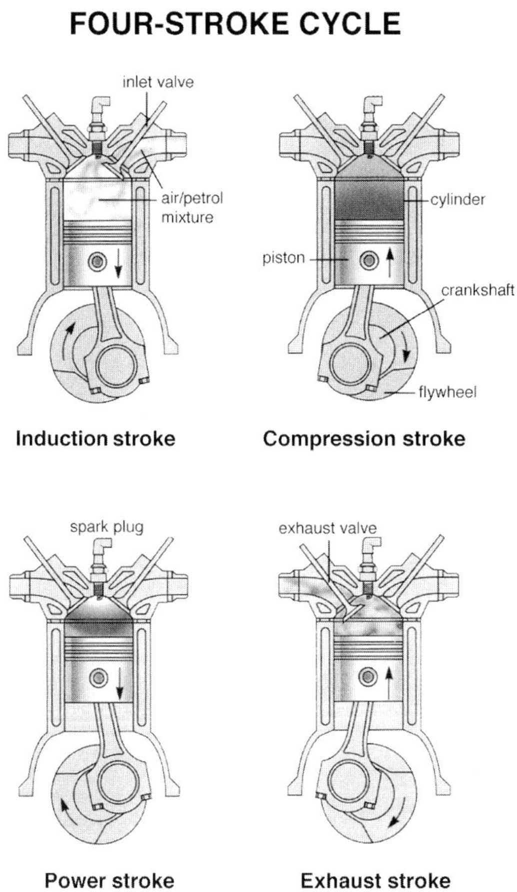

Induction stroke. The inlet valve opens. The piston moves down. A mixture of petrol and air is sucked into the cylinder.

Compression stroke. The inlet valve closes. The piston moves up. The mixture is squeezed into a small space above the piston.

Power stroke. A spark from the plug makes the mixture explode. Hot gases from the explosion force the piston down the cylinder.

Exhaust stroke. The outlet valve opens. The piston moves up. The burnt gases (exhaust gases) are forced out of the cylinder.

12.3 DIESEL ENGINES

Most trucks and buses use four-stroke diesel engines, as do some cars. A diesel engine does not need spark plugs. The compressed air gets so hot that the oil starts to burn as soon as it meets it. Diesel engines are less powerful than petrol engines but they use less fuel, and have less harmful exhaust gases.

12.4 ENGINE CONSTRUCTION

Cylinder Block. The biggest part of the engine is the cylinder block, which is also called an engine block(Fig. 12.2). The cylinder block is a large casting of metal that is drilled with holes to allow for the passage of lubricants and coolant through the block and provide spaces for movement of mechanical parts. The block contains the cylinders, which are round passageways fitted with pistons. The block houses or holds the major mechanical parts of the engine.

Fig. 12.2 The cylinder block is the foundation of the engine.

Cylinder Head. The cylinder head fits on top of the cylinder block to close off and seal the top of the cylinder. The combustion chamber is an area into which the air/fuel mixture is compressed and burned. The cylinder head contains all or most of the combustion chamber. The cylinder head also contains ports, which are passageways through which the air/fuel mixture enters and burned gases exit the cylinder.

Valve Train. A valve train is a series of parts used to open and close the intake and exhaust ports. A valve is a movable part that opens and closes a passageway. A camshaft controls the movement of the valves (Fig. 12.3), causing them to open and

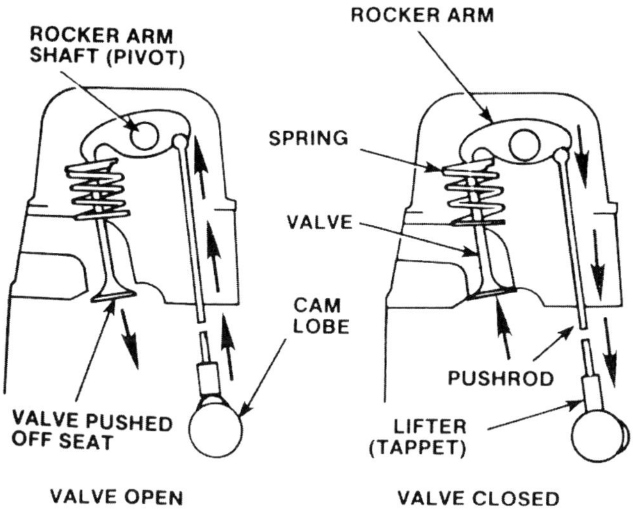

12.3 Valve train opens and closes the intake and exhaust ports

close at the proper time.

Piston. The burning of air and fuel occurs between the cylinder head and the top of the piston. The piston is a can-shaped part closely fitted inside the cylinder. In an engine with a four-stroke cycle, the piston has four strokes or movements: **intake**, **compression**, **power**, and **exhaust.** On the intake stroke, the piston moves downward, and a charge of air/fuel mixture is introduced into the cylinder. As the piston travels upward, the air/fuel mixture is compressed in preparation for burning. When the piston returns to the top of the cylinder, combustion occurs, and forces the piston downward on its power stroke. When it reciprocates, or moves upward again, the piston pushes the burned gases out of the cylinder.

Connecting Rods and Crankshaft The reciprocating motion

must be converted to rotary motion before it can drive the wheels of a vehicle.

As shown in Fig. 12.4, this conversion is achieved by linking

Fig. 12.4 The crankshaft changes reciprocating motion to rotary motion.

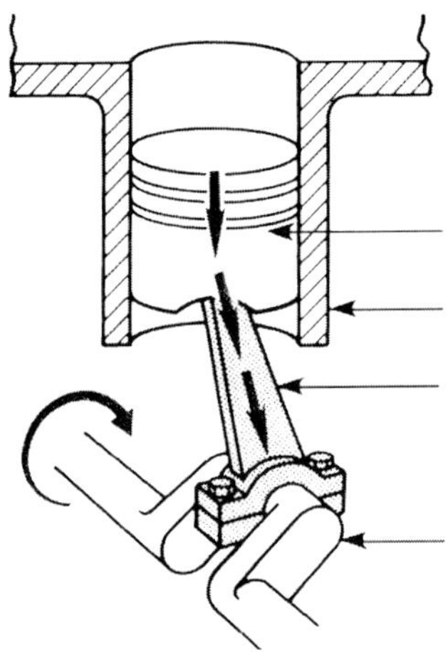

the piston to a crankshaft with a connecting rod. The upper end of the connecting rod moves with the piston as it moves up and down in the cylinder. The lower end of the connecting rod is attached to the crankshaft and moves in a circle. The end of the crankshaft is connected to the transmission to continue the power flow through the drivetrain and to the wheels.

Manifolds. A manifold is metal ductwork assembly used to direct the flow of gases to or from the combustion chambers. To separate manifolds, metal fixtures attach to the cylinder head (Fig. 12.5). The intake manifold delivers a mix of outside air and fuel vapor to the intake ports. The exhaust manifold mounts over the exhaust ports and carries exhaust gases away from the cylinders. The design of the manifolds has a critical effect on volumetric efficiency, or how easily an engine breathes. An engine that can gulp down more air per intake stroke burns more fuel, making more power than an engine that does not breathe as efficiently.

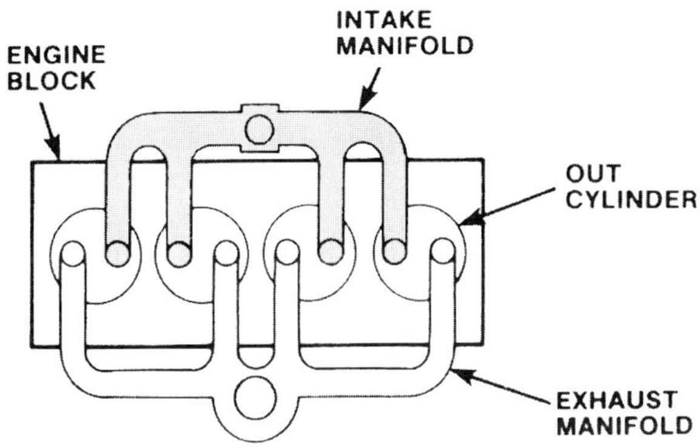

Fig. **12.5** Manifolds direct gases in and out of the combustion chambers.

Chapter 12

VOCABULARY

expand ; 넓히다. 팽창시키다.
experiment ; 실험
petrol ; 가솔린, 휘발유
ancestor ; 선조, 조상
ignite ; 불을 붙이다. 연소시키다.
explode ; 폭발(파열)시키다.
reliable ; 믿을 수 있는.
inlet valve ; 흡입밸브
suck ; 흡입하다, 빨아 들이다.
squeeze ; 압착하다, 죄다.
outlet valve ; 배기 밸브
induction stroke ; 흡입행정
compression stroke ; 압축행정
power stroke ; 폭발행정
exhaust stroke ; 배기행정
mixture ; 혼합, 혼합물
spark plug ; 스파크 플러그
harmful ; 유해한
lubricant ; 윤활유, 윤활제
coolant ; 냉각제
passageway ; 통로
combustion chamber ; 연소실
contain ; 담다, 포함하다.
exhaust port ; 배기구(排氣口)
can-shaped part ; 캔모양의 부품
preparation ; 준비
reciprocate ; 왕복운동을 시키다.
burned gas ; 연소가스
convert ; 변하게 하다, 전환하다.
attach ; 붙이다, 달다
transmission ; (자동차의)변속기
drivetrain ; 동력전달계통
manifold ; 분기관
ductwork ; (난방 등의)도관
assembly ; 조립부품
separate ; 분리하다, 갈라진
deliver ; (편지 등을)배달하다
vapor ; 증기
mount ; 앉히다, 설치하다.
critical ; 비평적인, 중대한
volumetric ; 용적[체적]측정의
efficiency ; 능률, 효율
breathe ; 숨쉬다. 호흡하다.
gulp ; 꿀꺽꿀꺽 마시다.

명언 / 격언

Never put off till tomorrow what can be done today.

— *Chesterfield*

James Watt

(1736 - 1819)

James Watt was born in Greenock, Scotland. He moved to Glasgow in 1754 to learn the trade of instrument maker. While he was employed on surveys for canals, he was also studying steam technology.

In 1763, while repairing a Newcomen engine, he found he could greatly improve the machine. His invention of the 'separate condenser' and the introduction of crank movements could make steam engines more efficient. After other improvements, he went into partnership with Matthew Boulton, and the new steam engine was manufactured at Birmingham in 1774. Several other inventions followed, including the double-acting engine, the centrifugal governor for automatic speed control, and the pressure gauge.

With this invention he provided one of the most essential components of early industrial revolution. The term horse-power was first used by him, and the power unit, the watt, is named in his honor.

13 Chapter

Computers

(컴퓨터)

13.1 INTRODUCTION

Electronic calculating machines were first made during World War II. The first working computer was buitl at manchester University, England, in 1948. From this the first commercial computer, the Ferranti Mark One, was developed in 1951 and filled a whole room. The first computers were large, because their memory circuits used ordinary wires and thousands of valves. Valves looked and glowed like light bulbs. Computers used them to store data by switching them on or off in a certain order.

Computers started to get smaller when transistors replaced valves. Transistors were no bigger than a pea and used much less power than valves. When scientists discovered how to pack thousands of transistors on a single chip, computers could be made even smaller.

Chapter 13

13.2 PERSONAL COMPUTER SYSTEM

A **system** is a set of related parts that operate together as a whole. You encounter systems every day. For example, your circulatory system moves blood through your body. This system

contains your heart, lungs, blood vessels, and so on. These parts are all related and work together to distribute energy throughout your body. Your stereo is also a system. It has parts like the turntable, amplifier, and speakers, which work together, in this case so you can hear music. A personal computer system,

Typical computer system Fig. 13.1

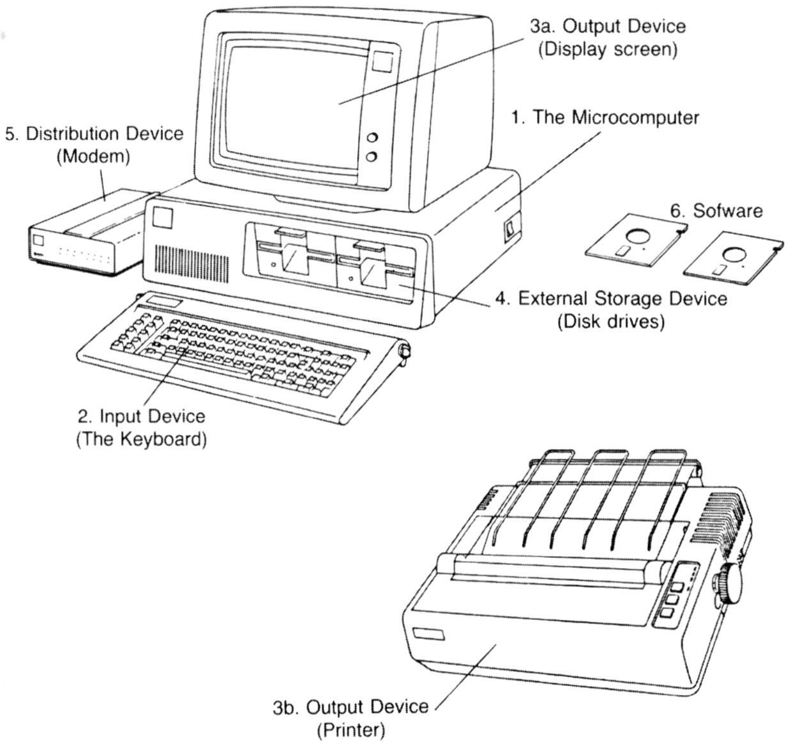

like these other systems, is also composed of several interacting parts (Fig. 13.1).

The central element in the system is the computer itself. Connected to the computer are a number of related parts, or **peripherals,** that you use to get data into and out of the computer and to store data and programs.

13.3 KEYBOARDS

You usually enter data that you want to process by typing it on the computer's keyboard. As you type, the information is fed to the central processing unit (CPU) and then either executed (if a command) or stored in memory (if data).

Keyboards vary in design and layout from computer to computer, but all have essentially the same types of keys. Fig. 13.2 shows a typical computer keyboard. The names of the keys and their location vary somewhat from keyboard to keyboard. Many keys have an **autorepeat** feature; that is, if you hold them down, they continue entering the key's character or repeating its function until you release the key.

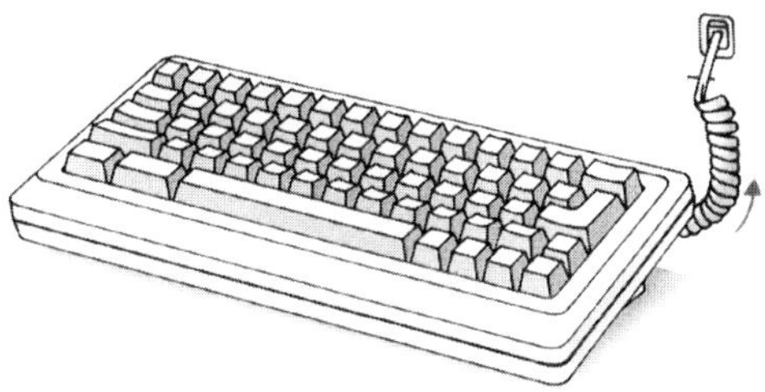

Fig. 13.2 Keyboard for typing data and programs into the computer. In some microcomputer, the keyboard and the computer are in the same plastic case.

13.4 MONITORS

The monitor attached to your PC gives you instant feed-back while you are working. A monitor is like a window into your computer' s memory. As you type text on the keyboard. It is entered into memory and echoed to the screen so you can see it. Since you spend most of your time looking at the monitor, its quality and capabilities are important. For example, some screens display just text characters, whereas others also display graphics. Some display an image in a single color, and others in multiple colors.

13.5 PRINTERS

The second most widely used output device is the printer (Fig. 13.3). Having a printer is essential if the data you process with your computer is to be filed or physically distributed to others. Until recently, you had to choose between an expensive printer that printed **letter-quality** characters like a type-writer and an inexpensive printer that created each character from a pattern of dots. These differences have narrowed, however, and today you have a wide variety of printers and prices to choose from. The quality of all types of printers has improved dramatically.

Hundreds of printers are available, and nearly all of them have unique features. There are, however, at least two ways to categorize them; by how they from characters and by how the characters are transferred to paper.

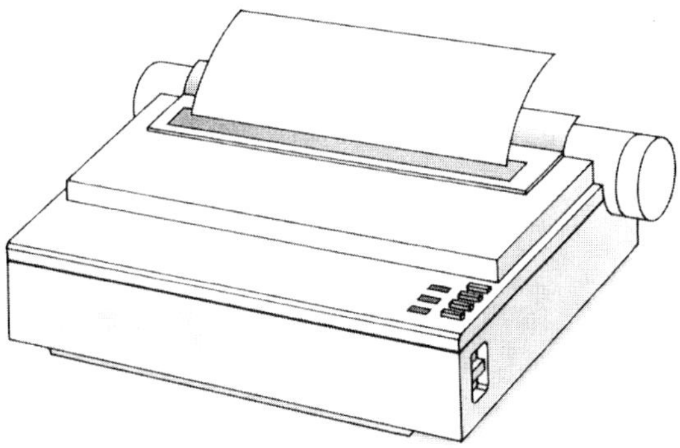

Fig. **13.3** Printer for printing on paper the data that you see on the screen. the copy printed on paper is called a hard copy.

13.6 FLOPPY DISKS AND DISK DRIVERS

Floppy disks. Floppy disks are available in three sizes: 8, $5\frac{1}{4}$, and $3\frac{1}{2}$ inches (Fig. 13.4). Each size works only with drives specifically designed to accept it.

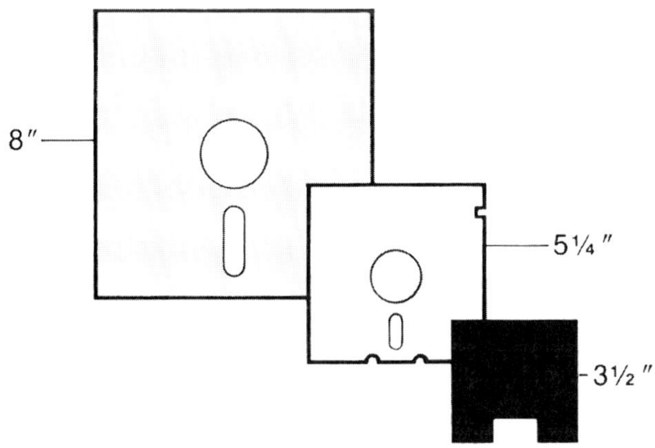

Fig. **13.4** Floppy disks

Floppy disk drive. The floppy disk drive (Fig. 13.5) is the device that the floppy disk is inserted into so that you can store data to and retrieve data from it. The floppy disk drive has two parts that you should be very familiar with: the drive' s slot and the driver' s light. The drive also has a light that tells you when it is operating so that you do not open the door until the drive stops spinning.

Floppy disk drive **Fig. 13.5**

13.7 HARD DISK DRIVES

Hard disk drives (Fig. 13.6), also called fixed disks. Instead of floppy disk, hard disk driver use rigid metal platters to store data. This allows them to store data more densely. This increased density plus the number of platters greatly increases their storage capacity. Hard disk drives generally provide 5, 10, 20, 40 and more megabytes of storage capacity. As Fig. 13.7 shows, a hard disk can store much more information than a floppy disk.

A hard disk drive spins at 3600 rpm, about ten times faster than a floppy disk drive, allowing data to be stored and retrieved faster. For example, an IBM PC-XT hard disk drive transfers data twenty times faster than a floppy disk drive.

13.6 Hard disks use metal platters instead of plastic disks to store data.

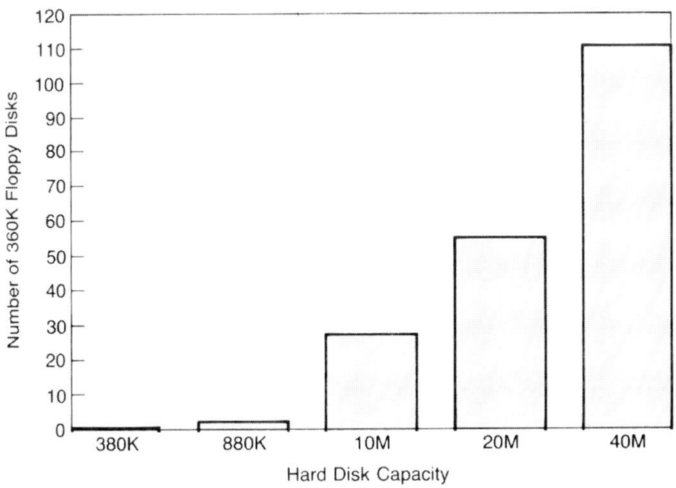

13.7 Hard disk can store much more information than floppy disks, as shown in this graph comparing the media. For example, a 10-megabyte hard disk drive can store the equivalent of twenty-seven 360 kilobyte floppy disks; a 20-megabyte hard disk, the equivalent of fifty-fivefloppy disks, and a 40-megabyte hard disk, the equivalent of 111 floppy disks.

Chapter 13

13.8 CD-ROM DISKS

CD-ROM disks, similar in concept to the compact disks (CDs) now popular in the music recording industry, can store so much information that it is measured in gigabytes. A small $4\frac{3}{4}$-inch CD-ROM can store up to 550 megabytes of data, or over $\frac{1}{2}$ gigabyte. This is equivalent to more than 1,500 floppy disks. Larger CD-ROMs can store up to 20 gigabytes and that limit will be exceeded long before you read this text.

13.9 MODEMS

A modem is a communications device that links computers connected by telephone lines. With modems, you can send data from your computer to another similarly equipped computer located anywhere you can reach by phone. This process in called **telecommunication.**

The reason you need a modem to telecommunicate is simple: The computer generates **digital signals**, but telephone lines are designed to carry **analog signals.** When you transmit a message, as Fig. 13.8 shows, the modem at the sending end converts the computer' s digital signals into analog signals (**modulation**) so that they can be transmitted effectively over telephone lines. The modem at the receiving end then converts the analog signals back into digital signals (**demodulation**) so that they can be used by that computer. The name *modem* derives from *modulate-demodulate.*

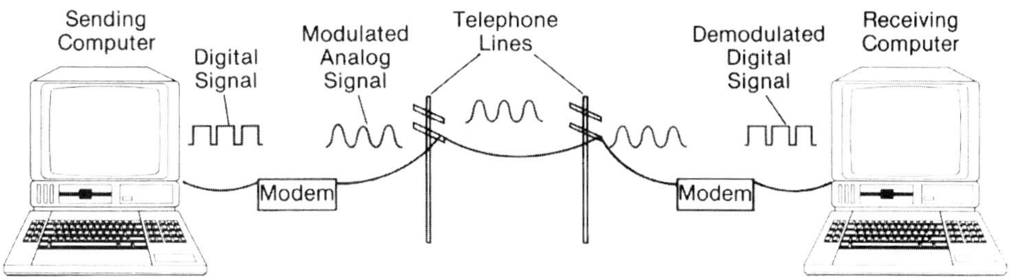

13.8 Modems are simply conversion devices. A modem at the sending end converts the digital signals from the computer into an analog signal that can be transmitted over the telephone wires. The modem at the receiving end converts the analog signal back into a digital signal.

Chapter 13

VOCABULARY

develop ; 발달시키다. 발전시키다.
memory ; 기억, 기억장치,
relate ; 관련(관계)시키다.
circulatory ; (혈액) 순환상의
blood vessel ; 혈관
distribute ; 분배하다, 배분하다.
interacting ; 상호작용 하는
densely ; 짙게, 빽빽이
megabyte ; 100만 byte
gigabyte ; 10억 바이트(정보단위)
execute ; 실행하다, 수행하다
layout ; 배치(도), 설계(법),
autorepeat ; 자동반복
release ; 풀어놓다.
echo ; 메아리 반향, 반복
capability ; 능력, 가능성
character ; 문자, 특성, 성격
output device ; 출력장치
distribute ; 분배하다, 배분하다.
variety ; 변화, 다양(성)
categorize ; 분류하다.
Floppy disk ; 플로피 디스크
insert ; 끼워넣다, 삽입하다.
retrieve ; 회복하다. 생각해내다.
platter ; (타원형의)큰 접시
peripheral ; 주위의, 주변장치의
information ; 통지, 정보, 지식
CPU (central processing unit) ; 중앙 처리장치
modem ; (컴퓨터의) 변복조장치
telecommunication ; (데이터, 음향, 영상 등의) 원격통신
generate ; 발생시키다, 생기게 하다.
digital signal ; 디지털 신호
analog signal ; 아날로그 신호
overall ; 전부의, 전체로서
shield ; 방패, 보호물
reflection ; 반사, 반영

명언 / 격언

First impression are most lasting.

— *Dikens*

Chapter 14

Technology of the Cars

(승용차)

14.1 INTRODUCTION

There are more than 400 million cars in the world today. Most have four wheels, though some have only three, and at least one has sixteen!

Saloon cars have a closed-in compartment for passengers, with a separate boot (trunk) for luggage. Coupes are saloons designed for two people. Estate cars (station wagons) are like vans with extra windows (Fig. 14.1). They have a door at the back and rear seats which fold down to give more luggage space.

Hatchbacks are like saloons, but with folding seats and a rear door like estates. Sports cars, are designed for fast driving. Most carry only two people. Some have fabric roofs which fold down. GT (Grand Touring) cars are saloons which perform like sports cars.

Chapter 14

Fig. 14.1 Typical station wagon body shapes

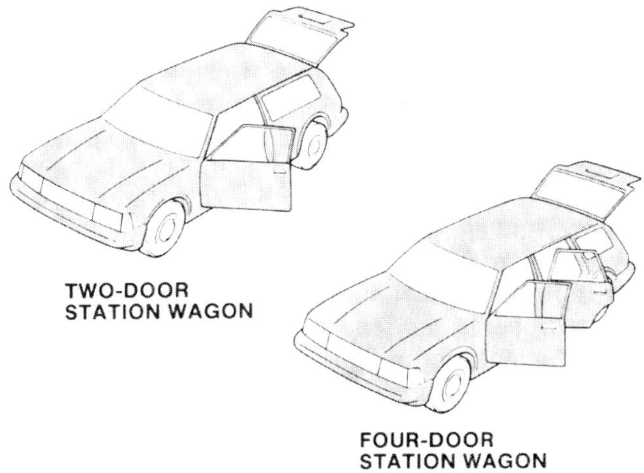

14.2 SAFETY

Modern cars are designed with safety in mind. The passenger compartment is built to resist collapse in an accident, while the front and rear sections are designed to crush up. This makes a collision less violent for the people inside.

There are seat-belts to stop passengers being thrown through the windscreen and special locks which stop the doors bursting open. The windows are made of safety glass which has no sharp edges when it breaks, and the steering column is collapsible in case the driver is thrown against it.

Some cars have bars in the doors to protect passengers in side-on collisions. And some have antilock brakes to make skids less likely.

14.3 FLASHBACK

The first petrol-driven cars were made in Germany in the 1880s. Passengers sat out in the open and, in Britain, someone had to walk in front waving a red flag. Early motoring was expensive. But, in 1907, Henry Ford in America started to produce a cheap reliable car, the Model T.

In the 1920s, a wide choice of cheap cars became available and the closed-in saloon body became popular. Since then, there have been many improvements but the basic design of the car has changed very little.

14.4 HOW A CAR WORKS

A car has over 20,000 parts, grouped into four main sections. The body holds the driver and passengers. Attached to this, there is an engine, a transmission system and a set of controls for the driver.

The engine is the heart of the car. It runs on petrol or diesel fuel and provides power to drive the wheels. The transmission system takes the power from the engine to the wheels. For the driver, there is a steeringwheel to point the car in the right direction, an accelerator to make it go faster and brakes to slow it down. The driver may also have a clutch and a gearlever for changing gear, though in some cars this is done automatically.

14.5 CONSTRUCTION OF CARS

Engine. This provides the power to drive the car. It is normally an internal combustion engine, burning petrol or diesel fuel. You push down the accelerator to make it go faster. Some cars have the engine at the rear.

Transmission and clutch. Most cars have four of five forward gears and one reverse gear. The low gears, called first and second, are used for staring off and driving up steep hills.

The higher gears are used for normal driving. To use the clutch, you press a foot pedal. It disconnects the engine from the gearbox while you change gear.

Driveshafts. These are metal rods (one either side) which transmit the turning force from the engine and the gearbox to the wheels.

Larger cars often have rear wheel drive (Fig. 14.2A). Most small cars are front wheel driven like this one (Fig. 14.2B). The back wheels just roll along. In some cars, all four wheels are driven (Fig. 14.2C).

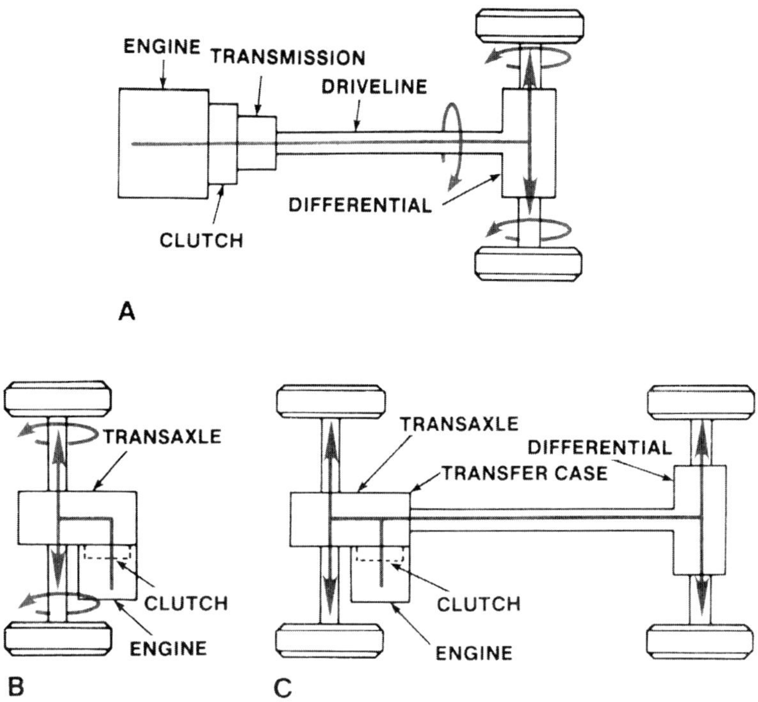

14.2 Driveshafts :
(A) rear-wheel drive,
(B) front-wheel drive,
(C) four-wheel drive

Suspension System. This joins each wheel to the car body. It allows the wheels to go up and down over bumps while giving the passengers a smooth ride.

Most suspensions use coil springs. They have dampers (shock absorbers) to stop the car bouncing up and down.

Steering System. The car is steered by turning the front wheels in the direction you want to go. The steering-wheel is connected to the wheels by a steering-box and two rods which move from side to side (Fig. 14.3).

Common steering systems *Fig.* **14.3**

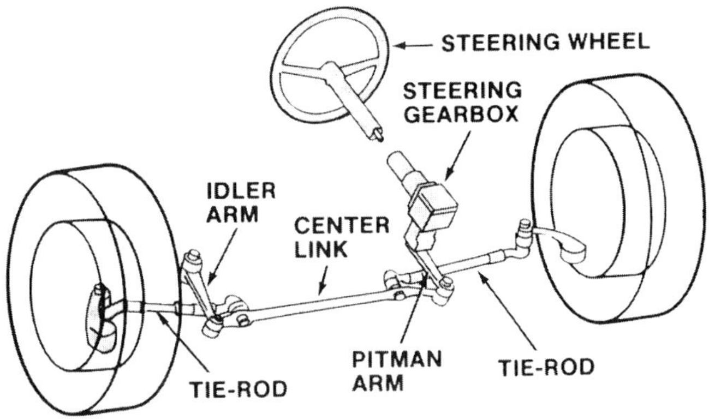

Brakes System. These are attached to the wheels and suspension. You use them to slow or stop the car. The footbrake pedal works on all four wheels. The handbrake works on two wheels and is used for parking. There are two kinds of brakes: disc brakes and drum brakes. Most cars have disc brakes at the front and drum brakes at the back.

Cooling system. When the engine is running it gets very hot. To stop it overheating, the cooling system pumps water around the engine and then through the radiator.

The hot water in the radiator is cooled either by a rush of air passing through as the car drives along or by a fan when the car is not moving.

Electrical System. A battery provides electricity for the starter motor and spark-plugs when starting the engine. It also powers the lights and the radio when the engine is not on. When the engine has been started, it drives an alternator.

This supplies the electricity for the lights, plugs and other equipment. It also recharges the battery.

14.6 INSTRUMENTS FOR THE DRIVER

Speedometer tells you the speed of the car in km/h or mph.

Odometer tells you the total distance travelled by the car in kilometers or miles.

Tachometer (rev. counter) tells you the engine speed in revolutions per minute (rpm).

Chapter 14

VOCABULARY

compartment ; 칸막이, 구획, 칸막이방
passenger ; 승객, 여객
separate ; 가르다, 분리하다
boot ; 자동차의 트렁크
skid ; (자동차 바퀴의) 미끄럼
flashback ; 플래쉬백 (과거장면 으로의 순간적인 전환)
transmission ; 변속기
accelerator ; 가속장치
automatically ; 자동적으로
windscreen ; (자동차 앞의)유리
burst ; (폭탄 등이) 파열하다
antilock brake ; 미끄럼방지 브레크
luggage ; 수하물
saloon ; 특별 객차, 세단형 자동차
fold ; 접다, 포개다
fabric ; 직물, 천
collapse ; (건물 등이) 무너지다.
crush ; 충돌, 격돌

명언 / 격언

Prevention is better than cure.

— *Dickens*

Biography of Henry Ford

(Born July 30, 1863 - Died April 7, 1947)

Henry Ford was born in Wayne County, Michigan on July 30, 1863. His father was born in Ireland and came to America in 1847. Young Ford showed an early interest in mechanics. At age 15, he constructed his first steam engine. In 1893, Ford built a small one-cylinder gasoline internal combustion engine. From there, he mounted an engine on a frame fitted with four bicycle wheels; his first automobile.

In August, 1899, Ford formed the Detroit Automobile Company, becoming its chief engineer. On June 16, 1903, Henry organized the Ford Motor Company, capitalized at $150,000, of which only $28,000 in cash was paidin. In 1906, he became president and controlling owner.

His aim was mass consumption by mass ownership of automobiles. Ford' s moving assembly line led to high-speed manufacturing and ultimately sparked the 20th century industrial revolution. Henry Ford died on April 7, 1947, at the age of 83.

15

Aircraft Chapter

(항공기)

15.1 INTRODUCTION

Aircraft is the name we use for flying machines. Most have wings to keep them in the air but helicopters and hot-air balloons are really aircraft as well.

During the 19th century, short gliding flights were made in craft built by Sir George Cayley in England and Otto Lilienthal in Germany. However, it was not until 1903 that powered flight was achieved by the Wright brothers in the USA (Fig 15.1).

Fig 15.1 The Wrights' first successful plane

Chapter 15

In 1909, Louis Bleriot flew across the English Channel. Ten year later, Alcock and Brown made the first non-spot flight across the Atlantic.

Passenger flying developed after World War I. During the 1920s, passengers often flew aboard mail planes, sometimes with the mailbags on their laps! By the late 1930s, flying was a much more luxurious affair. People could travel between Europe and the Far East aboard large flying boats which took off and landed on water.

The first jet aircraft, the Heinkel HE178, flew in 1939. However, jets were developed mainly after World War II. The first jet airliner, the De Havilland Comet, entered service in 1952. Today, nearly all long-distance international travel is by jet.

15.2. SHAPES

Most aircraft have a central body called a fuselage, with wings near the middle and horizontal stabilizer and vertical stabilizer at back (Fig. 15.2)

Straight wings work best for carrying heavy loads at low speed, but swept-back wings give a better airflow for fast flying. Some military jets, such as the Panavia Tornado, have `swing wings' which swing further back for high-speed flight. Some aircraft do not have a tailplane. Instead, the wings form a triangular shape, called a delta, which goes all the way to the back.

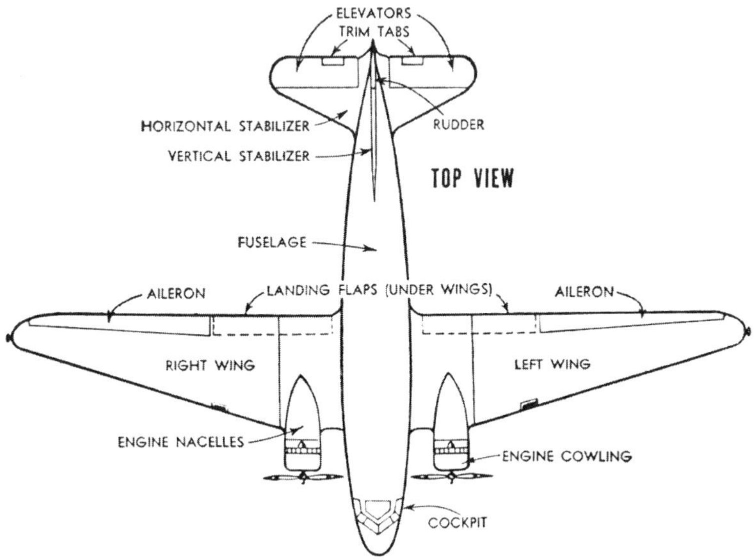

Fig. **15.2** Basic parts and control surfaces of an airplane

Concorde is like this (Fig. 15.3). Delta wings are good for high-speed flight but do not perform well at low speed. When Concorde lands, its wings are set at such a steep angle that the nose has to be lowered to give the pilot a forward view. A few aircraft have their wings at the back and their 'tailplane' at the front. This called a canard arrangement.

Fig. **15.3** A concorde

15.3 POWER

Most modern aircraft use jet engines in one form or another. Even where a propeller is fitted, the power may come from a turbo-prop engine which is based on the jet. Until the 1950s, most aircraft had propellers turned by piston engines. Some small aircraft still do. These engines work in the same basic way as a motor car engine.

15.4 THE CHANGING FACE OF DESIGN

Concorde was designed in the 1960s. Nowadays, airlines are more interested in economy than speed, and other airliners travel at less than half Concorde's speed. Using computers, designers have developed wings which slip more easily through the air, and engines which are quieter and burn their fuel more efficiently. A`jumbo jet' can carry four times as many passengers as Concorde using only the same amount of fuel.

Computers are an important part of a modern airliner. The autopilot is a computer which can navigate and fly the aircraft for most of its journey. On some aircraft, the pilot does not directly control the plane. Instead, the pilot' s controls send instructions to a computer and the computer works out the best way to fly the plane.

15.5 GLOSSARY FOR AIRCRAFT

Monoplane. An aircraft with a single set of wings. Most modern aircraft are monoplanes.

Biplane. An aircraft with two sets of wings, one above the other. Most aircraft built before the mid-1920s were like this. Some modern aerobatic aircraft are biplanes.

Amphibian. A plane which is able to land eight on water or on land. Such planes have pontoons or boat bodies and also retractable wheels.

Glider. An aircraft heavier than air, similar to an airplane but without a power plant.

Chapter 15

VOCABULARY

balloon ; 기구, 풍선
powered flight ; 동력비행
mail plane ; 우편 비행기
luxurious ; 사치스러운, 호화로운
take off ; 이륙하다.
airliner ; (대형의) 정기 여객±®
fuselage ; (비행기의)동체, 기체
horizontal ; 수평의, 가로의
stabilizer ; 안정장치(선박, 항공기의)
pontoon ; (tn상비행기의)플로트
retractable ; 신축성있는
power plant ; 동력발생장치
bachelor ; 미혼남자
triangular ; 삼각형의
delta ; 델타 (그리스어 알파벳의 네 번째 글자)
efficiently ; 능률적인, 효과있는
autopilot ; 자동조정장치
navigate ; (바다,하늘을)항해하다.
monoplane ; 단엽 비행기
biplane ; 복엽 비행기
amphibian ; 수륙양용의 비행기
swept-back wing ; (후퇴각을 가진) 날개
airflow ; 기류
tailplane ; 수평꼬리날개

명언 / 격언

The important thing in life is to have a great aim, and the determination to attain it.

— *Goethe*

The Wright Brothers:

Orville(1871-1948) and Wilbur(1867-1912)

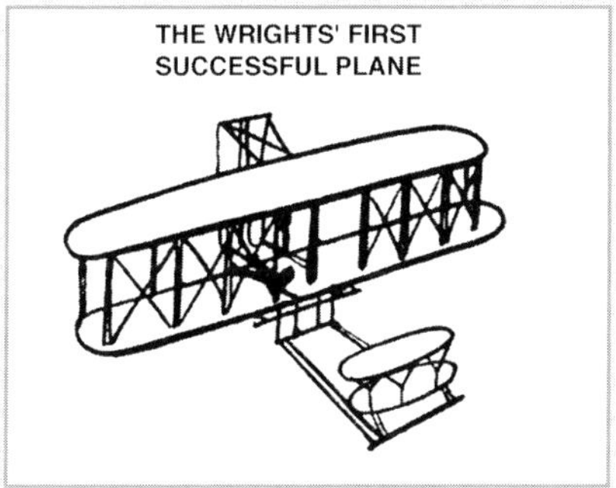

Orville was born in Ohio, and his brother Wilbur was born in Indiana. Mainly self-educated. Built and designed printing machinery and bicycles. Did experiments on flying in their free time.

From 1899 to 1902, built gliders (planes without engines) to find the principes of controlling movement in the air. Found that they needed three controls : for going right and left, for going up and down, and for stopping the plane from turning upside down.

In 1903 in North Carolina, made first powered and controlled flight. By 1905, made half-four flights with first practical airplane. Began making planes for U.S. Army and France. Showed their planes in Germany and Italy.

Both brothers were modest, hard-working bachelors interested only in airplanes. Wilbur died in 1912. Orville continued to work with airplanes until 1948.

APPENDICES

1. Greek Letters
2. Standard SI prefixes
3. Metric Equivalents
4. Cardinal Numerals and Ordinal Numerals
5. Mathematical Symbols
6. Abbreviations for Use on Drawing
7. Technical Terms

1. Greek Letters

(그리스 문자)

A	α	Alpha	N	ν	Nu
B	β	Beta	Ξ	ξ	Xi
Γ	γ	Gamma	O	o	Omicron
Δ	δ	Delta	Π	π	Pi
E	ε	Epsilon	P	ρ	Rho
Z	ζ	Zeta	Σ	σ	Sigma
H	η	Eta	T	τ	Tau
Θ	θ	Theta	Υ	υ	Upsilon
I	ι	Iota	Φ	φ	Phi
K	κ	Kappa	X	χ	Chi
Λ	λ	Lambda	Ψ	ψ	Psi
M	μ	Mu	Ω	ω	Omega

2. Standard SI Prefixes

(표준 SI단위의 접두어)

Name		Symbol	Factor
exa	(엑사)	E	1 000 000 000 000 000 000 = 10^{18}
peta	(페타)	P	1 000 000 000 000 000 = 10^{15}
tera	(테라)	T	1 000 000 000 000 = 10^{12}
giga	(기가)	G	1 000 000 000 = 10^{9}
mega	(메가)	M	1 000 000 = 10^{6}
kilo	(킬로)	k	1 000 = 10^{3}
hecto	(헥토)	h	100 = 10^{2}
deka	(데카)	da	10 = 10^{1}
deci	(데시)	d	0.1 = 10^{-1}
centi	(센티)	c	0.01 = 10^{-2}
milli	(밀리)	m	0.001 = 10^{-3}
micro	(마이크로)	μ	0. 000 001 = 10^{-6}
nano	(나노)	n	0. 000 000 001 = 10^{-9}
pico	(피코)	p	0. 000 000 000 001 = 10^{-12}
femto	(펨토)	f	0. 000 000 000 000 001 = 10^{-15}
atto	(아토)	a	0. 000 000 000 000 000 001 = 10^{-18}

3. Metric Equivalents

(미터식 환산)

Length

U.S. to Metric	Metric to U.S.
1 inch = 2.540 centimeters	1 millimeter = .039 inch
1 foot = .305 meter	1 centimeter = .394 inch
1 yard = .914 meter	1 meter = 3.281 feet or 1.094 yards
1 mile = 1.609 kilometers	1 kilometer = .621 mile

Area

U.S. to Metric	Metric to U.S.
1 inch^2 = 6.451 centimeter^2	1 millimeter^2 = .00155 inch^2
1 foot^2 = .093 meter^2	1 centimeter^2 = .155 inch^2
1 yard^2 = .836 meter^2	1 meter^2 = 10.764 foot^2 or 1.196 yard^2
1 acre^2 = 4,046.873 meter^2	1 kilometer^2 = .386 mile^2 or 247.04 acre^2

Volume

U.S. to Metric	Metric to U.S.
1 inch^3 = 16.387 centimeter^3	1 centimeter^3 = .061 inch^3
1 foot^3 = .028 meter^3	1 meter^3 = 35.314 foot^3 or 1.308 yard^3
1 yard^3 = .764 meter^3	1 liter = .2642 gallons
1 quart = .946 liter	1 liter = 1.057 quarts
1 gallon = .003785 meter^3	1 meter^3 = 264.02 gallons

Weight

U.S. to Metric	Metric to U.S.
1 ounce = 28.349 grams	1 gram = .035 ounce
1 pound = .454 kilogram	1 kilogram = 2.205 pounds
1 ton = .907 metric ton	1 metric ton = 1.102 tons

Velocity

U.S. to Metric	Metric to U.S.
1 foot/second = .305 meter/second	1 meter/second = 3.281 feet/second
1 mile/hour = .447 meter/second	1 kilometer/hour = .621 mile/second

Acceleration

U.S. to Metric	Metric to U.S.
1 inch/second^2 = .0254 meter/second^2	1 meter/second^2 = 3.278 feet/second^2
1 foot/second^2 = .305 meter/second^2	

Force

N(newton) = basic unit of force, kg-m/s^2 . A mass of one kilogram (1kg) exerts a gravitational force of 9.8N(theoretically 9.80665N) at mean sea level.

4. Cardinal Numerals and Ordinal Numerals
(기수와 서수)

기 수			서 수	
명 칭	아라비아 숫자	로마숫자	명 칭	간약형
naught, zero	0		zeroth	
one	1	I	first	1st
two	2	II	second	2nd, 2d
three	3	III	third	3rd, 3d
four	4	IV	fourth	4th
five	5	V	fifth	5th
six	6	VI	sixth	6th
seven	7	VII	seventh	7th
eight	8	VIII	eighth	8th
nine	9	IX	ninth	9th
ten	10	X	tenth	10th
eleven	11	XI	eleventh	11th
twelve	12	XII	twelfth	12th
thirteen	13	XIII	thirteenth	13th
fourteen	14	XIV	fourteenth	14th
fifteen	15	XV	fifteenth	15th
sixteen	16	XVI	sixteenth	16th
seventeen	17	XVII	seventeenth	17th
eighteen	18	XVIII	eighteenth	18th
nineteen	19	XIX	nineteenth	19th
twenty	20	XX	twentieth	20th
thirty	30	XXX	thirtieth	30th
forty	40	XL	fortieth	40th
fifty	50	L	fiftieth	50th
sixty	60	LX	sixtieth	60th
seventy	70	LXX	seventieth	70th
eighty	80	LXXX	eightieth	80th
ninety	90	XC	ninetieth	90th
one hundred	100	C	one hundredth	100th
one thousand	1000	M	one thousandth	1000th
ten thousand	10000	X	ten thousandth	10000th

5. Mathematical Symbols

(수학기호)

기 호	읽는 법	해 석
+	plus	플러스
−	minus	마이너스
±	plus or minus	플러스 또는 마이너스
∓	minus or plus	마이너스 또는 플러스
×	multiplied by	곱하기
÷	divided by	나누기
=	is equal to	같다
≒	approximately equals	거의 같다, 약…
>	is larger than	…보다 큰
≫	is much greater than	…보다 훨씬 더 큰
≧	is greater than or equal to	…보다 크거나 같다.
≯	is not greater than	…보다 크지 않다.
<	is less than	…보다 작은
≪	is much less than	…보다 훨씬 작은
≦	is less than or equal to	…보다 작거나 같다.
≮	is not less than	…보다 작지 않은
⊥	is perpendicular to	…에 수직
≠	is not equal to	같지 않다.
≡	is identically equal to	동일(합동)
∠	angle	각
∥	paralell to	…와 평행하다.
⊿	increment of	…의 증가
∼	the difference between	…의 차
	(예) a∼b는	
	the difference between a and b	a와 b의 차라고 읽는다.
∝	varies as	…에 비례한다.
	(예) a∝b는 a varies as b	a는 b에 비례한다.
:	the ratio of	…의 비
∴	therefore	그러므로
( )	brackets or parenthesis	소괄호
{ }	braces	중괄호
[ ]	angular brackets	대괄호
%	percent	퍼센트

기호	읽 는 법
$\cdots{}'$	prime or dash
$\cdots{}''$	second or two dash
$\cdots{}'''$	third or three dash
	(예) a', a'', a''' 는 각각 a prime, a second, a third라 읽는다.
½	a half or one half
⅓	a third or one third
⅔	two thirds
¼	a quarter (one quarter, one fourth도 가끔 쓰인다.)
¾	three quarters (three fourth도 가끔 쓰인다.)
3⅛	three and one eighth
$^{113}/_{300}$	one hundred thirteen over three hundred
38.72	thirty eight decimal (point) seven two
0.45	decimal (point) naught four five
4.9	four point nine recurring
3.0326	three point naught three two six, two six recurring
	[주의] 0(제로)와 영문자 O(오우)는 같은 모양이므로 naught라고 읽는대신 O(오우)라 읽어도 좋다.
$(8+6\frac{5}{6}-3.88\times4)\div2\frac{1}{2}$	Eight plus six and five-sixths minus three decimal (point) eight eight multiplied by four, all divided by two and half.
a^2	a square or a squared
a^3	a cube or a cubed
b^4	b to the fourth
$\sqrt{518}$	the square root of five hundred and eighteen
	[주의] the square root of 대신에 the root of 또는 the radical of 라고도 한다.
$\sqrt[3]{930}$	the cubic root of nine hundred and thirty
$\int$	integral

6. Abbreviations for Use on Drawings

(제도에 사용되는 약어)

ALY	alloy	GAL	gallon
AC	alternating current	HEX	hexagon
AL	aluminum	HS	high-speed
AWG	American wire gage	HP	horsepower
AMP	ampere	HR	hour
ANT.	antenna	HYD	hydraulic
ARC/W	arc weld	(˝)IN	inch
ASSY	assembly	ID	inside diameter
AUTO	automaic	LAB	laboratory
AUX	auxiliary	LH	left hand
BLK	block	LG	length
BC	bolt circle	MACH	machine
BHP	brake horsepower	MFG	manufacturing
BRZG	brazing	MAX	maximum
BH	Brinell hardness	MPH	miles per hour
BTU	British thermal unit	MM	millimeter
BRZ	bronze	MIN	minimum
CAP SCR	cap screw	(´)MIN	minute
CS	cast steel	NO.	number
CSTG	casting	OD	outside diameter
CL	center line	PC	pitch circle
CG	center of gravity	PD	pitch diameter
CHAM	chamfer	LB	pound
CW	clockwise	PSI	pounds per square inch
CONST	constant	RAD	radial
CBORE	counterbore	R	radius
CCW	counter clockwise	RPM	revolutions per minute
CDRILL	counterdrill	RH	right hand
CPUNCH	counterpunch	RD	root diameter
CSK	cuntersink	SEC	second
(°)DEG	degree	SPG	spring
DSGN	design	STD	standard
DIA	diameter	STL	steel
DP	diametral pitch	SUM	summary
EA	each	SYS	system
ENG	engine	TECH	technical
(´)FT	feet	TOL	tolerance
FIG.	figure	VS	versus
FL	floor	VOL	volume
FOC	focus	W	watt

7. Technical terms

"The beginning of wistom is to call things by their right names."
- Chinese provers

지식은 사물의 이름을 정확히 부르는 것으로부터 시작된다.
-중국 속담

acme thread(n) 애크미 나사, 29° 사다리꼴나사

addendum(n) 어덴덤 (이끝높이), 기어에 있어서 피치원으로부터 위쪽의 이(齒) 높이.

allen screw(n) 머리에 6각 홈이 있는 캡나사

allowance(n)허용차; 끼워맞춤에 있어서 축과 구멍과의 치수차 .

alloy(n)합금; 2종 이상의 금속을 서로 융합시켜 만든 금속.

aluminum(n) 알루미늄(원소기호 Al). 비중 2.7, 융점 600℃, 은백색이고 가벼우전연성이 좋아 가공하기 쉬운 금속이다.

angle steel(n) 단면이 90°의 각을 이루는 형강.

anneal(v) 애닐링 하다.; 열을 가하고 천천히 식히므로서 취성을 없애고 연성을 증가시키는 것을 말한다.

arc-welding(n) 아크용접법; 전기아크에 의한 열을 이용하여 금속을 용접하는 방법.

babbit metal(n) 바비트 메탈; 주석 70~90%에 소량의 안티몬과 구리를 첨가한 베어링 합금.

bar stock. 봉재(棒材) 막대모양으로 생긴 재료

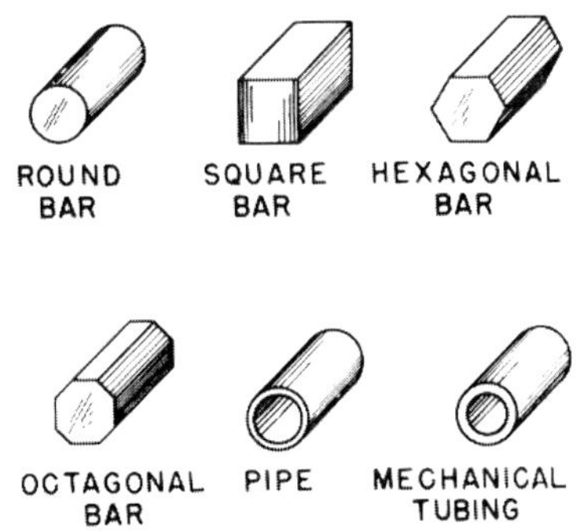

bearing(n) 베어링; 회전축을 지지하는 역할을 하는 부재를 베어링이라 한다.

bevel gear(n) 베벨 기어; 삿갓모양의 차(車에) 이를 붙인 기어로서 2축이 어느

각도를 이루고 있는 경우의 전동에 사용된다.

bolt circle(n) 볼트원; 볼트의 중심을 지나는 원을 말한다.

bore(v) 보어; 보링밀(공구의 명칭)로 구멍을 확대시키.

boss(n) 보스; 주조에서 원통형 돌출부를 보스라고 한다.

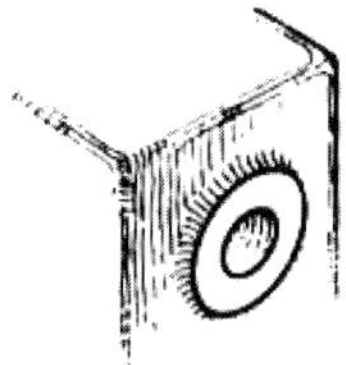

BOSS

brass(n) 황동; 구리에 아연을 첨가한 황색의 합금.

brazing(n) 납땝; 저융점의 합금을 녹혀 금속에 접합시키는 방법.

broach(n) 브로우치; 봉의 외주에 많은 상사형의 칼날을 축에 따라서 치수 순으로 배열한 절삭공구. 주로 키홈을 파는데 사용된다.

bronze(n) 청동; 동에 주석을 첨가한 합금. 베어링,동상,주화 등을 만드는데 사용된다.

buff(v) 버프; 금속의 표면을 연마하기 위하여 원형의 천 또는 피혁을 봉합한 것.

burnish(v) 비니시; 연마하다.

burr(v) 비어; 금속재료를 절단하거나 구멍뚫기하는 경우 그 갓부분에 생기는 벗겨진것과 같은 흠집.

bushing(n) 버니싱; 보링 또는 연마한 구멍에 다듬질 강구 또는 주판알 모양의 공구를 압입하여 표면의 凹凸을 압과 다듬질하는 특수가공방법을 말한다.

calipers(n) 캘리퍼스; 공작물 두께나 환봉의 지름을 측정하는 공구.

cam(n) 캠; 판 또는 원통의 곡선단 또는 홈을 만든 1종의 차(車로)서 대개 구동절로 회전하고 종동절에 주기적 운동을 준다.

carburizing(n) 침탄(浸炭)

case-harden 표면을 경화시킨다. 표면을 담금질하다.

castellated shaft 스플라인 샤프트

casting(n) 주조; 금속을 고온가열하여 용해하고 용해된 금속을 주형에 흘려넣어 냉각해서 성형하는 방법.

cast iron(n) 주철 탄소 1.7～6.8%를 포함하는 철의 합금. 실제로 사용되고 있는 것은 탄소 2.5～4.5%로서 Si, Mn, P, S 등도 포함되어 있다.

center drill(n) 센터드릴; 센터구멍을 뚫는데 사용하는 드릴. 그림에서 보는 바와 같이 보통 카운터싱크 드릴과 함께 사용된다.

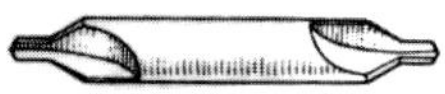

COMBINED DRILL
& C SINK

chamfer(n) 챔퍼; 모서리를 뗀 사면.

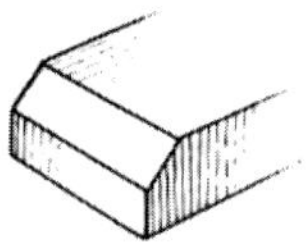

CHAMFER

chaser(n) 체이서; (빗형바이트) 나사를 절삭하는 공구의 일종; 여러개의 나사산 모양을 한 절삭 공구.

cheek(n) 몰딩에 사용되는 2쪽 플래스크의 중간부분.

chill(v) 칠; 주조표면을 급냉시킴으로써 경하게 하는 것을 말한다.

chip(v) 칩; 금속을 절삭하다.

chuck(n) 공작물을 고정하는 공구로서 특히 선반의 것은 주축에 고정하고 조우와 공작물과의 마찰로 공작물을 물리도록 만들어져있다.

coin(v) 코이닝; 압인가공, 조각형 1조의 다이(die) 사이에 재료에 압력을 가하여 표면에 凹凸을 만드는 가공법.

cold-rolling(n) 상온 압연 또는 냉간 압연; 상온에서 압연하는 것.

collar(n) 칼라; 축이나 관등에 끼워넣는 연강제의 둥근테.

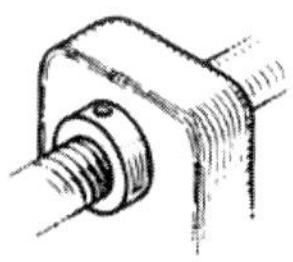

COLLAR

colorharden(v) 표면을 경화시키다.

cope(n) 모딩에 사용되는 플라스크의 위쪽 부분.

core(v)코어; 주조에서 구멍을 형성하다.

coreprint(n) 코어프린터; 코어를 주형내에 유지하기 위하여 만든 돌출부.

cotter pin(n) 코터핀; 탈출 방지핀, 코터가 빠지지 않도록 끼우는 핀.

cotter pin(n)

counterbore(v) 구멍의 끝을 원통형으로 넓게 하는 것.

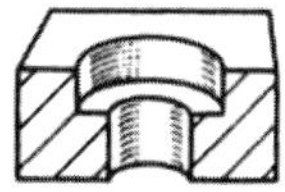

COUNTERBORE

countersink(v) 구멍의 끝을 원추형으로 넓게 만드는 것.

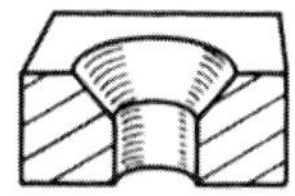

COUNTERSINK

crown(n) 크라운; 벨트차로부터 벨트가 이탈하지 않도록 볼트차의 림(rim)을 볼록하게 만든 부분을 말한다.

cylinder(n) ①실린더; 유체를 밀폐한 원통형의 용기. 증기기관, 내연기관, 공기압축기관, 펌프 등 왕복기관의 중요부.

dedendum(n) 디덴덤; 기어의치에 있어서 피치원에서부터 아랫까지의 치높이.

development(n) 입체의 표면을 전개하여 1 평면상에 넓혀서 그림 도형.

diametral pitch(n) 지름피치, DP=치수/피치원의 경(D의 단위 in).

die(n) 다이; 숫나사 깍는 공구. 둥근 다이와 각다이가 있다.

die casting(n) 다이 캐스팅; 정밀한 금형(다이형)을 사용하여 자동 또는 수동으로 주탕하고 압력을 가하여 정밀도가 높은 주조물을 주조하는 방법.

die stamping(n) 다이로 금속판재를 성형하는 공정

dog(n) 돌리개, 캐리어.

dowel(n) 자리맞춤; 목형을 2분할 또는 그 이상으로 분할하여 만드는 경우 이들이 어긋나지 않도록 맞춤면에 끼우도록 만든 작은 핀.

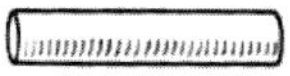

DOWEL

draft(n) 통풍; 연료의 연소에 필요한 바람을 로(爐)나 보일러등에 보내는 것. 인공통풍과 자연통풍이 있다. ② 발출구배, 주조작업에서 주형속의 모형을 빼내기 쉽도록 목형 수직면에 약간의 구배를 붙인것.

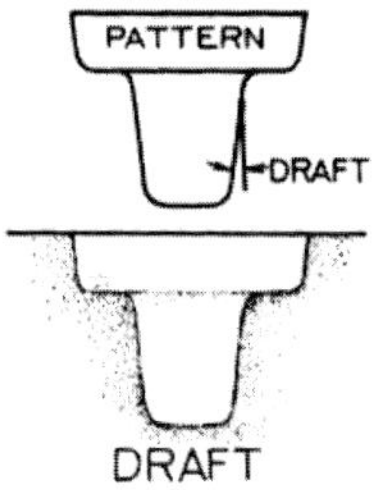

DRAFT

drag(n) 아랫주형상자; 분할되어 있는 하측이 되는 부분.

draw(v) 금속을 잡아당겨 늘리다.

drill(v) 원통형 구멍을 뚫다.

drill press(n) 드릴프레스, 드릴링머시인; 보올반은 독일어로 Bohr Machine에서 나온말. 공작물은 테이블 위에 놓고 드릴에 회전과 이송을 주어 공작물에 구멍을 뚫는 기계.

drop hammer(n) 낙하햄머

face(v) 선반에서 회전축심에 90° 되는 면을 가공하는 것을 페이싱이라고 한다.

feather key(n) 미끄럼 키이. =sliding key.

file(v) 줄; 양질의 강편 또는 공구강편에 날을 만들어 당금질한 손공구

fillet(n) 필렛; 갈고리형으로 구부러진 부분이나 갈고리형으로 접속된 부분의 내각.

fin(n) 휜; ① 냉각 또는 절연면적을 넓게 하기 위하여 붙이는 지느러미 모양의 부분.

fixture(n); 픽스츄어, 고정구(固定具); 부품가공 때 공작물을 가공할 수 있도록 기계에 고정하는 장치.

flange(n) 프렌지, 테; 보강 또는 접속을 위하여 전(全)둘레에 붙인 돌출부. 외향플랜지와 내향플랜지가 있다.

flash(n) ; fin과 같다.

flask(n) 주조상자 (=molding box)

flute(n) 홈 ; 드릴이나 리이머 등의 날의 홈부를 말한다.

forge(v) 단조하다.

galvanize(v) 전기도금하다.

gasket(n) 가스킷; 얇은 판상의 패킹.

gate(n) 湯口; 주조작업에서 용융한 금속을 주형에 부어넣는 입구.

graduate(v) 동사) ①눈금을 새기다. ② 등급을 붙이다.

grind(v) ①연삭하다. ②가루로 갈다.

harden(v) 경도를 증대시키다.

heat-treat(v) 열처리하다.

interchangeable(adj.) 서로 교환할 수 있는

journal(n) 저어널; 축경(軸頸) 전동축에 있어서 베어링속에 들어가 잇는 부분을 특히 저어널이라 한다.

kerf(n) 쇠톱에 의하여 만들어진 홈

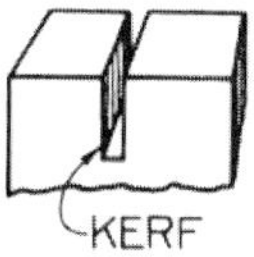

key(n) ① 키이; 기어, 벨트차등에 축을 고정하는데 사용하는 가늘고 긴 쐐기형의 것.

keyseat(n) 축에 있는 키홈

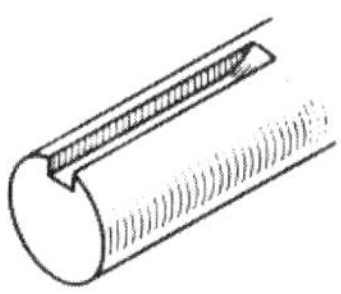

KEYSEAT

keyway(n) 허브이 있는 키홈

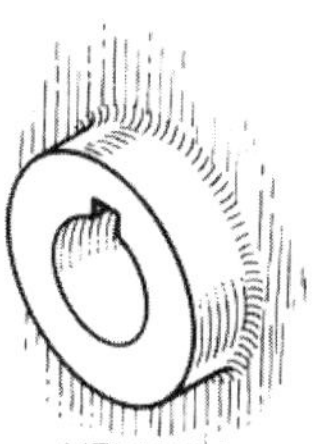

KEYWAY

knurling(n) 널링; 공구나 계기류 등의 손잡이 부분이 미끌어지지 않도록 톱니모양의 돌기부를 붙이는 공작법

lap(v) ①겹침,중첩. 동사) 겹치다. 중첩되다.

lathe(n) 선반 (旋盤)

lug(n) ①돌출부, 突起点.

malleable casting(n) 가단주철로 만든 주물

mill(v) ①제조공장.제조소. (동사)빻다

mold(n) 주형용융금속(탕)을 흘러 넣어 주물을 만드는 형.

MS(n) 연강(mild steel)

neck(v) 원통둘레에 넥크홈을 파다.

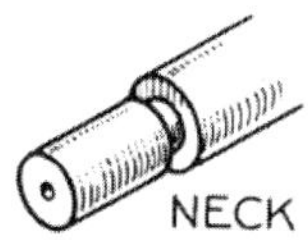

packing(n) 패킹; 관의 이음매, 기타 접합부에서 기체가 새어나오지 않도록 패킹제를 끼워 두는 것. 박판상의 패킹을 가스킷(gasket)이라고 부른다.

pattern(n) ①목형(木型); 주형을 만드는 원형이 되는 나무로 만든 형.

peen(v) 해머의 두부. 보통 사용되는 평편한 면의 반대의 일단. 동사) 쳐서 늘리다. (해머의 머리로)굽히다. 치다.

pickling(n) 산세척법(酸洗滌法) ; 단조품 등을 산속에 담가 산의 작용으로 스케일(scale) 등을 제거하는 법

pinion(n) 피니언, 대소가 맞물리는 두 개의 기어중 작은 쪽의 기어. 특히 랙과 맞물리는 기어를 말한다.

pitch circle(n) 피치원

plane(v) 프래너에 의하여 평탄하게 깍다.

planish(v)

plate(v) ①판(板), 평면판.

polish(v) ①정밀연마. ②광택, 광택 촉진

제 동사)연마하다. 광택을 내다.

profile(v) 윤곽(측면에서) 윤곽을 그리다

punch(v) ①펀치; 금긋기용구의 1종(구멍 등을) 타발하다.

quench(v) 경화시키기 위하여 열을 가한 금속을 물 또는 기름에 담그다.

rack(n) 랙; [기어] 곧은 봉에 치철삭을한 것으로서 피니언(작은 기어)과 맞물려 회전운동을 직선운동으로 바꾸는데 사용된다.

ream(v) (구멍을) 넓히다.(=widen)

relief(n) 릴리이프; 기계다름질에서 작업이 곤란한 장소에 오목부를 만들고 다듬질을 피하는 것.

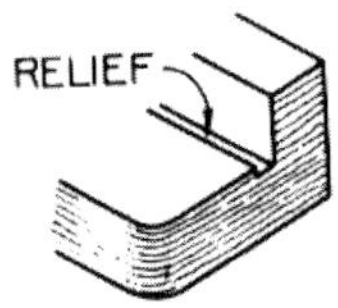

rib(n) 리브; 판상 또는 박육의 부분을 보강하기 위하여 붙이는 골재.

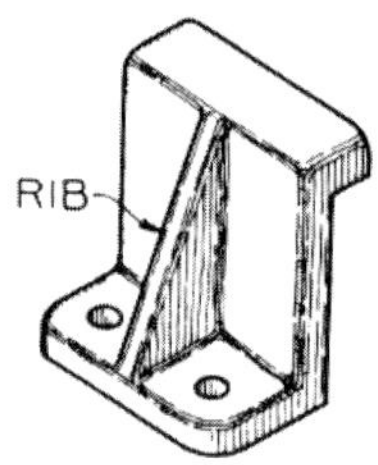

rivet(v) 리벳; 환봉에 머리가 붙은 것. 판재나 형재를 영구적으로 체결화하는 기계요소의 하나이다. (동사)리벳 체결을 하다.

round(n) 두 표면이 만나는 부분을 둥글게 만든 것

SAE 미국자동차 공업협회(Society of Automotive Engineering)

sandblast(v) 표면을 깨끗하게 하기 위하여 압축공기로 모래를 고속으로 불어넣다.

scleroscope(n) 금 속의 경도를 측정하는 기기(器機)

scrape(v) 스크레이퍼로 금속을 깍아내다.

shape(v) (동사)형성하다.

shear(v) (동사) 전단하다.
전단기(=shearing machine)

sherardize(v)

shim(n) 두 금속 사이의 간격을 조정하기 위한 얇은 금속편

solder(v) 납으로 때우다.

spin(v)

spline(n) ①스플라인(=spline shaft).

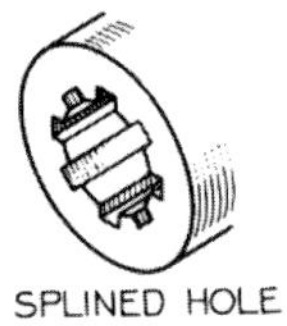

spot-face(v)

steel casting(n) 주조 (=cast steel).

steel shape 보통 열간전조에 의하여 만들어진 상업용 강철재

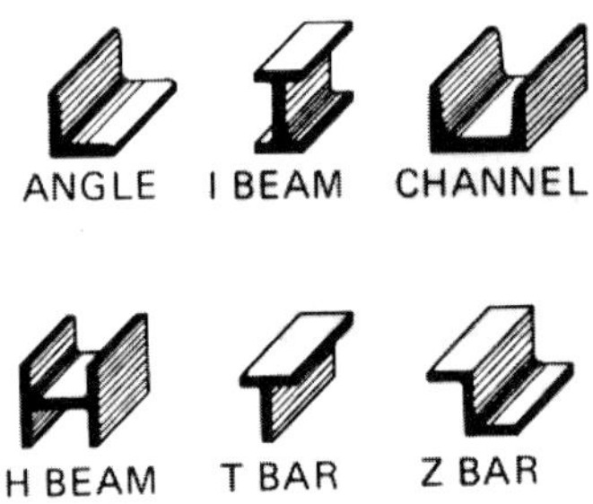

swage(n) 닳구어진 쇠를 어떤 형태로 두드려 만들기 위하여 대장간에서 사용하는 특수한 형태의 공구를 말한다.

sweat(v) 납을 사용하여 두 금속을 접합시키는 것을 말한다.

tap(v) (비교적 작은) 암나사를 깍다.

taper pin(n) 테이퍼 핀(=tappered pin); 테이퍼가 붙은 핀.

TAPER PIN

taper reamer(n) 테이퍼 리이머; 테이퍼 구멍을 다듬질하는데 사용하는 리이머.

temper(v) (강철등을) 탬퍼링하다. 원하는 경도를 얻기 위하여 경화된 강을 다시 가열하다.

tin(n) 주석; 원조기호 Sn. 비중 7.3, 융점 23℃ 공기중에서 안정하고 가공하기 쉬운 백색의 금속. 합석판을 만드는 외에도 청동, 베어링 합금 등의 중요 성분이 된다.

tolerance(n) 공차; 공작품의 가공에서 최대치수와 최소치수와의 차이. 즉 허용되는 오차의 범위를 말한다.

trepan(v)

tumble(v)

turn(v) 중심선에 평행한 방향으로 원통표면을 깍다.

twist drill(n) 드릴프레스에 사용되는 드릴

upset(v)

web(n) 웨브; 홈형강, 기타 이에 유사한으로서 중앙복부를 말한다.

weld(v) ①용접(=welding); 금속합금을 점성상태 또는 반용융상태로 하여 접합하는 방법. 전기용접, 가스용접 등으로 대별된다.

Woodruff key(n) 반달키이; 반달모양의 특수한 키이를 말한다.

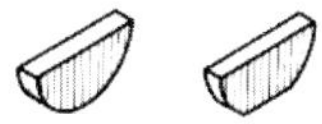

WOODRUFF KEYS

wrought iron(n) 연철; 연강과 같이 반용융상태에서 제련한 연철로서 C 0.2%이하를 함유한다.

Index

accelerator · 185
ACCIDENTS · 103
Achilles' tendon · 106
Acme thread · 117
Addendum · 131
adjustable wrench · 14
air compressor · 158
Aircraft · 193
airliner · 196
airlines · 196
ALIGNED SECTIONS · 44
Alloy steels · 61
Aluminium · 61
Aluminium alloys · 63
Amphibian · 197
analog signals · 180
Angle of thread · 115
annular gears · 133
antilock brakes · 184
aprons · 96, 99
arc welding · 93
Archimedes · 161
autopilot · 196
autorepeat · 175
Backlash · 132
ball peen hammer · 18
balsa wood · 154
bar · 75
barrier cream · 109
Bearing housing · 63
BENDING · 81
Bevel gears · 68, 134
Bilateral Tolerance · 49
Biplane · 197
blacksmith · 83
BLANKING · 80
boiler suit · 104
box wrench · 13
brake drums · 82
Brakes System · 188
brass · 53, 62
BROKEN-OUT SECTIONS · 43
Bronze · 62
buoyant force · 154
buttress thread · 118
Calipers · 27
CAP SCREW · 121
carbon · 59
cast iron · 59, 60
Casting · 71
CD-ROM disks · 180
center punch · 21
chip · 173
chisel · 19
Circular pitch · 131
Clearance · 132
Clearance fit · 50
clutch · 185
cold rolling · 76
combination wrench · 14
compound gear train · 139
Compression stroke · 164
compressive strength · 55
compressive stress · 143
Computers · 173
Concorde · 195

conductors · 53
Connecting Rods · 167
CONVENTIONAL BREAKS · 45
Cooling system · 188
Copper · 53, 61
Crankshaft · 167
Crest · 115
CUFFS · 106
Cylinder Block · 165
Cylinder Head · 166
Dedendum · 131
demodulation · 180
density · 152
Diametral pitch · 131
digital signals · 180
discs · 111
DISK DRIVERS · 177
double shear · 149
double threads · 116
DRAWING · 82
Drilling · 88
Driveshafts · 186
drivetrain · 168
Ductility · 55
ductwork · 168
Edison · 142
elastic limit · 146
Electrical System · 188
electrode · 97
EUREKA · 161
Exhaust stroke · 164
external thread · 113
EYE PROTECTION · 108
EYEBOLT · 120
Facing · 87
factor of safety · 147
fasteners · 113
Feeler gages · 26
Ferrous Metals · 59
file · 20
first-angle · 35
FLASHBACK · 185
FLAT KEY · 125
Floppy disk drive · 178
Floppy disks · 177
flux · 98
FOOT PROTECTION · 109
footwear · 96, 109
Forging · 71
FOUR-STROKE CYCLE · 164
FULL SECTIONS · 38
FUSE CLIP · 81
fuselage · 194
gas welding · 93
gas-welded joint · 95
gauntlets · 99
gear train · 137
Gears · 129
gigabytes · 180
Glider · 197
gloves · 96
Goggle · 20
GRINDING · 90
hacksaw · 22
HALF SECTIONS · 40
hammers · 18
Hand forging · 83
HAND PROTECTION · 109
HARD DISK DRIVES · 178
Hardness · 55
HEAD PROTECTION · 107
helicopters · 193
Henry Ford · 185, 191
High-carbon steel · 60
HOOK' S LAW · 144
hot rolling · 76
hot-air balloons · 193
hydraulic lift · 158
hydraulic press · 157
hydraulic press · 78, 79
hydraulic system · 151
idler gears · 138
impact strength · 57
Induction stroke · 164
ingots · 75
Injection moulding · 85
inside caliper · 27
insulator · 53
Interference fit · 50
Internal gears · 134
INTERNAL GEARS · 133
internal thread · 113

iron · 59
James Watt · 171
jaws · 17
jet engines · 196
keyboard · 175
Keys · 125
knuckle thread · 117
lathe · 87
Lead · 61, 115
Limit Dimensioning · 50
LOCK WASHERS · 123
MACHINE SCREW · 122
Maching · 71
machining · 86
Major diameter · 114
Malleability · 55
Manifolds · 168
Medium-carbon steel · 60
megabytes · 180
metric thread · 117
micrometer · 28
Mild steel · 61
Milling · 88
Minor diameter · 114
modem · 180
Module · 131
monitor · 176
Monoplane · 197
Motorcycle engine · 68
Multiple threads · 115
muscles · 111
nippers · 17
Non-ferrous Metals · 59
Non-metallic Materials · 59
Odometer · 189
OFFSET SECTIONS · 39
open-end wrench · 13
oxy-acetylene · 94
pascal · 155
PASCAL' S PRINCIPLE · 157
peripherals · 175
permissible stress · 147
petrol engines · 163
petrol-driven cars · 185
Phillips screw · 12
phonograph · 142
PICTURE HOOK · 79
PIERCING · 80
pin punch · 21
Pinion · 131, 132
Pins · 126
PIPE CLIP · 81
Piston · 167
Pitch · 114
Pitch circle · 131
Plastics Molding · 71
Pliers · 17
pneumatic system · 151
Power stroke · 164
Pressing · 71
PRESSING · 78
printer · 176
Punches · 21
rack · 132
ratchet handle · 15
REAMING · 89
REMOVED SECTIONS · 42
resistance welding · 93
REVOLVED SECTIONS · 41
RIBS IN SECTION · 43
RISER · 73
Rolling · 71
rolling mill · 78
Root · 115
RULER · 25
RUNNER · 73
Saloon · 183
saucepans · 82
SCREW TREAD · 113
screwdriver · 11
Section lining · 48
SECTIONING · 37
service station · 158
SETSCREWS · 122
SHAPES · 194
shear strength · 55
shear stress · 148
SHEARING · 148
simple gear train · 137
single thread · 115
skids · 184
slag · 98

Snap rings · 124
socket wrench · 15
sparkplug · 26, 63
specific gravity · 152
Speedometer · 189
Spot welding · 100
spur gear · 68, 129, 130
SQUARE KEY · 125
square thread · 117
stabilizer · 194
starting punch · 22
station wagons · 183
steam engine · 163
steel scale · 25
Steering System · 187
steeringwheel · 185
strain · 143
Strength · 54
Strength of Materials · 143
STRESS · 143
Stress-strain graph · 145
STUD BOLT · 120
Suspension System · 187
Tachometer · 189
tap · 71
TAP BOLT · 119
taper pins · 126
TAPPING SCREWS · 123
telecommunication · 180
tensile strength · 55, 146
tensile stress · 143
THIN SECTIONS · 45
third-angle · 35
Thomas Young · 145
Tin · 61
TOLERANCE · 48
torch · 94
torque wrench · 16
Toughness · 54
Transistors · 173
Transition fit · 50
transmission · 168
triple threads · 116
TROUGH BOLT · 119
turbo-prop engine · 196
TURNBUCKLE · 121
turning · 87
U-BOLT · 120
Unified thread · 117
Unilateral Tolerance · 49
Valve Train · 166
velocity ratio · 137
vernier caliper · 29
visible outline · 34
visor · 99
WASHER · 79
water systems · 151
Welding · 93
Whole depth · 132
Wire feeler gages · 26
Woodruff key · 126
worm · 136
worm and wheel · 136
worm gear · 136
WRENCHES · 12
Wright brothers · 193
yield stress · 146
YOUNG' S MODULUS · 144
Zinc · 61

알기쉬운 **기계기술영어**

2000년 2월 29일 1쇄 발행
2002년 1월 30일 2쇄 발행
2004년 3월 10일 3쇄 발행
2007년 2월 25일 4쇄 발행
2009년 3월 10일 5쇄 발행
2012년 1월 30일 6쇄 발행

지은이 | 정 태 환
발행인 | 김 병 계
발행처 | 도서출판 보문당
등록 | 1990년 6월 15일 (제1-1066호)
주소 | 서울시 마포구 신수동 448-6 우)121-110
전화 | 704-7025 / 704-0635 FAX | 704-2324
홈페이지 | www.bomoondang.co.kr
전자우편 | bmdpub@naver.com

정가 12,000 원

ISBN 978-89-8413-037-1